D1031800

Mobile Robots

Mobile Robots

Navigation, Control and Sensing, Surface Robots
and AUVs

Second Edition

Gerald Cook
George Mason University

Feitian Zhang
George Mason University

WILEY

For general information on our other products and services or for technical support, please contact our Customer Care Department within the United States at (800) 762-2974, outside the United States at (317) 572-3993 or fax (317) 572-4002.

Wiley also publishes its books in a variety of electronic formats. Some content that appears in print may not be available in electronic formats. For more information about Wiley products, visit our web site at www.wiley.com.

Library of Congress Cataloging-in-Publication Data is available.

hardback: 9781119534785

10 9 8 7 6 5 4 3 2 1

Gerald Cook

To my heavenly Father for leading me to a vocation that has brought me a lifetime of joy and happiness.

To my wife, Nancy Anne, for her encouragement and support of all my endeavors throughout my career.

To my two adult sons, Bo and Ford, for their continued encouragement and interest in my work.

To my mother, Rose Boyer Cook, who as a single parent of four provided so abundantly for our needs.

Feitian Zhang

To my wife, Mi Zhou, and our children, Andy and Lisa, for bringing me strength and joy that encourage me throughout my career.

To my parents, Guangbo Zhang and Dongmei He, for their unconditional love and support all the time.

Contents

Preface

A number of experiences and acquaintances have contributed to this project. The Countermine Branch of the Science Division of the Night Vision Electronic Sensors Directorate (NVESD), United States Army played a particularly important role through its sponsorship of related research. This research effort had as its objective the detection and geo-registration of landmines through the use of vehicular mounted sensors. The nature of the problem required that a broad set of tools be brought to bear. These required tools included a vehicle model, sensor models, coordinate transformations, navigation, state estimation, probabilistic decision making, and others. Much of the required technology had previously existed. The contribution here was to bring together these particular bodies of knowledge and combine them so as to meet the objectives. This led to several interesting years of interaction with NVESD and other researchers in this area of applied research.

Afterwards, it was realized that the work could be cast in a more general framework, leading to a set of notes for a second-year graduate course in Mobile Robots. A course on modern control and one on random processes are the required prerequisites. This course was taught several times at George Mason University, and numerous revisions and additions resulted as well as a set of problems at the end of each chapter. Finally, the notes were organized more formally with the result being the first edition of this book.

I would like to express my appreciation to some of the individuals who have influenced and encouraged me in the writing of this book. These include Kelly Sherbondy, my research sponsor at NVESD, former colleague Guy Beale, department chairman Andre Manitius, former student Patrick Kreidl, industrial associate Bill Pettus, collaborator at the Naval Research Laboratory Jay Oaks, former students Smriti Kansal and Shwetha Jakkidi who were part of the NVESD project, and the many other students who have attended my classes and provided me with inspiration over the years.

Gerald Cook

The major addition to the second edition of this book includes modeling and control of autonomous underwater vehicles (AUVs), which exhibits unique complex three-dimensional dynamics. The materials are mainly based on my PhD research project on design, modeling, and control of a novel underwater vehicle named gliding robotic fish that is essentially a hybrid of underwater glider and robotic fish. The research, sponsored by National Science Foundation (NSF), aimed to develop an autonomous platform for aquatic environmental monitoring through fundamental understanding and effective control of gliding robotic fish, which eventually led to generalized modeling and control approaches for AUVs written in this book. I would like to acknowledge and thank my PhD advisor Xiaobo Tan, my collaborators Hassan Khalil at Michigan State University and Fumin Zhang at Georgia Institute of Technology for their enormous support and insightful guidance in the research project, and my colleague Gerald Cook for motivating and encouraging me in co-writing the second edition of this book.

Feitian Zhang

The following is a suggested schedule for teaching a one-semester course from this book.
1) Kinematic Models for Mobile Robots: 0.5 weeks.
2) Mobile Robot Control: 1.5 weeks.
3) Robot Attitude: 1.0 week.
4) Robot Navigation: 2.0 weeks.
5) Application of Kalman Filtering: 1.5 weeks.
6) Remote Sensing: 1.5 weeks.
7) Target Tracking Including Multiple Targets with Multiple Sensors: 1.0 week.
8) Obstacle Mapping and Its Application to Robot Navigation: 1.0 week.
9) Operating a Robotic Manipulator: 1.0 week.
10) Remote Sensing via UAVs: 0.5 weeks.
11) Dynamics Modeling of AUVs: 1.0 week.
12) Control of AUVs: 1.5 week.

It is hoped that this book will also serve as a useful reference to those working in related areas. Because of the overriding objective described in the title of the book, the topics cut across traditional curricular boundaries to bring together material from several engineering disciplines. As a result, the book could be used for a course taught within electrical engineering, mechanical engineering, aerospace engineering, or possibly others. We would like to acknowledge here that MATLAB® is a registered trademark of The MathWorks, Inc. Also, please note, two of the videos referred to in Appendix A can be viewed at https://www.wiley.com/en-us/Mobile+Robots%3A+Navigation%2C+Control+and+Remote+Sensing%2C+2nd+Edition-p-9781119534785.

About the Authors

Gerald Cook, ScD, is the Earle C. Williams Professor Emeritus of Electrical Engineering and past chairman of Electrical and Computer Engineering at George Mason University. He was previously Chairman of Electrical and Biomedical Engineering at Vanderbilt University and before that, Professor of Electrical Engineering at the University of Virginia. He is a Life Fellow of the Institute of Electrical and Electronics Engineers (IEEE), a former president of the IEEE Industrial Electronics Society and a former Editor in Chief of the IEEE Transactions on Industrial Electronics.

Feitian Zhang, PhD, is an Assistant Professor in the Department of Electrical and Computer Engineering at George Mason University. He received the Bachelor's and Master's degrees in Automatic Control from Harbin Institute of Technology in China, and the PhD degree in Electrical and Computer Engineering from Michigan State University. He was a Postdoctoral Research Associate in the Department of Aerospace Engineering at the University of Maryland prior to joining Mason. His research interests include robotics, control, artificial intelligence, and underwater vehicles.

Introduction

I wish to take this opportunity to express my appreciation to Dr. Feitian Zhang for joining with me as Co-Author in developing this second edition of *Mobile Roots*. He has demonstrated a high level of knowledge and skill in the area of autonomous underwater robots (AUVs) and adds a new dimension to the book with this contribution. It has been a pleasure working together on this project.

Mobile robots, as the name implies, have the ability to move around. They may travel on the ground, on the surface of bodies of water, under water, and in the air. This is in contrast with fixed-base robotic manipulators that are more commonplace in manufacturing operations such as automobile assembly, aircraft assembly, electronic parts assembly, welding, spray painting, and others. Fixed-base robotic manipulators are typically programmed to perform repetitive tasks with perhaps limited use of sensors, whereas mobile robots are typically less structured in their operation and likely to use more sensors.

As a mobile robot performs its tasks, it is important for its supervisor to maintain knowledge of its location and orientation. Only then can the sensed information be accurately reported and fully exploited. Thus navigation is essential. Navigation is also required in the process of directing the mobile robot to a specified destination. Along with navigation is the need for stable and efficient control strategies. The navigation and control operations must work together hand-in-hand. Once the mobile robot has reached its destination, the sensors can acquire the needed data and either store it for future transfer or report it immediately to the next level up. Thus, there is a whole system of functions required for effective use of mobile robots.

Mobile robots may be operated in a variety of different modes. One of these is the teleoperated mode in which a supervisor provides some of the instructions. Here sensors including cameras provide information from the robot to the supervisor that enables him or her to assess the situation and decide on the next

Mobile Robots: Navigation, Control and Sensing, Surface Robots and AUVs,
Second Edition. Gerald Cook and Feitian Zhang.
© 2020 by The Institute of Electrical and Electronics Engineers, Inc.
Published 2020 by John Wiley & Sons, Inc.

course of action. The supervision may be very complete, leaving no decision making to the robot, or it may be at a high level only, leaving details to be worked out by algorithms residing on the robot. Some examples of this type of operation are the Mars rovers and the walking robots that descended down into the volcano on Mount Saint Helens in the state of Washington. Additional applications include the handling of hazardous materials such as nuclear waste or explosives and the search in war operations for explosives such as landmines. Other examples are unmanned air vehicles (UAVs) and AUVs that can be used for reconnaissance operations. The trajectory may be prespecified with the provision for intervention and redirection as the circumstances dictate.

One of the more interesting stories involving a teleoperated mobile robot took place in Prince William County, Virginia in the nineties. The police had a suspect cornered in an apartment house and decided that since he was armed they would send in their mobile robot. It was a tracked vehicle with a camera, an articulated manipulator, and a stun gun. Under the direction of a supervisor the robot was able to climb the stairs, open the apartment door, open a closet door, lift a pile of clothes off the suspect, and then stun him so that he could be apprehended. This served a very useful purpose and alleviated the need for the police officers to subject themselves to risk of injury or death.

Another possible mode is autonomous operation. Here the robot operates without external inputs except those inputs obtained through its sensors. Often there is a random element to the motion with sensors for collision avoidance and/or signal seeking. One example of this type of operation was the miniature solar-powered lawn mowers at the CIA in Langley, Virginia. These mobile robots were the size of a dinner plate and had razor sharp blades. The courtyard in which they worked was quite smooth with well-defined boundaries. Each robot could move in a random direction until hitting an obstacle at which time it switched to a new direction. Another example of this autonomous robotic behavior is a swimming-pool cleaner. This device moves about the pool sucking up any debris on the bottom of the pool and causing it to be pumped into the filtration system. The motion of the mobile robot seems to be somewhat random with the walls of the pool providing a natural boundary. Similar devices exist for vacuuming homes or offices.

A very exciting and recent example of an underwater semi-autonomous vehicle was the crossing of the Atlantic Ocean, from the coast of New Jersey to the coast of Spain, by the deep-sea glider Scarlet. This 8-ft long, 135 lb, unmanned vehicle was the product of a research team at Rutgers University and Teledyne Webb Research. The voyage took 221 days, extended over 4,600 miles, and provided data on the water temperature and salinity as a function of depth. The glider was powered by a battery that alternately pumped water out of the front portion of the vehicle to cause it to rise and took on water to cause it to dive. The battery could also be shifted forward or backward to modify the weight distribution and thereby adjust the glide angle. As the glider dove or climbed, its hydrodynamic wings gave it forward motion in much the same manner as that of a toy airplane glider dropped from a second floor window. It was equipped with a rudder for steering. Normally it traveled down to a depth of 600 ft below the surface of the ocean and then up to within 60 ft of the surface. A few times per day it would surface to get a GPS fix on its position, make radio contact with its supervisor and obtain a new way-point to head toward. Apparently the vehicle was equipped with an inertial measurement device that would provide heading information while underwater. (*Washington Post*, Tuesday, December 15, 2009, health and science Section pages E1 and E6.) As was mentioned, an important application of AUVs such as this is data collection of variables such as water temperature and salinity as a function of location, including depth.

Examples of mobile robots in manufacturing facilities include wheeled vehicles used for material transfer from one work station to another. Here a line painted on the floor may designate the path for the mobile robot to follow. Optical sensors sense the boundaries of the line and give commands to the steering system to cause the mobile robot to follow along the track. Schemes such as this can also be used for mobile robots whose assignment is to perform inventory checks or security checks in a large facility such as a warehouse. Here the path for the mobile robot is specified and the sensors acquire and store the required information as the robot makes its rounds.

There are two basic types of steering used by mobile robots operating on the ground. For both of these types of steering, the mobile robot may have one or two front wheels. One type is front-wheel steering much like that of an automobile. This type of steering presents interesting challenges to the controller, because it yields a nonzero turning radius. This radius is limited by the length of the robot and the maximum steering angle.

The other type of steering involves independent wheel control for each side. By rotating the left and right wheels in opposite directions at the same speed, the robot can be made to turn while in place, i.e., at a zero turning radius. Tracked vehicles use this same type of differential-drive steering strategy, there often referred to as skid steering.

Examples of mobile robots also include, as we mentioned earlier, AUVs such as underwater gliders, whose diverse applications range from oil/gas exploration and environmental monitoring to search and rescue and national harbor security. Due to the complex interaction between surrounding fluid and AUVs, hydrodynamics play an important role in determining vehicle dynamics which exhibits high nonlinearity. In addition, AUVs operate in open water environments typically in a truly three-dimensional trajectory. Therefore, it is essential to establish the dynamic model of AUVs and further investigate how to control AUV's dynamic motions given the unique propulsion and steering mechanisms such as buoyancy adjustment and control surfaces (e.g., a rudder or an elevator).

The objectives of this book are to serve as a textbook for a one-semester graduate course on wheeled surface robots as well as AUVs and also to provide a useful reference for one interested in these fields. The book presumes knowledge of modern control and random processes. Exercises are included with each chapter. Prior facility with digital simulation of dynamic systems is very helpful but may be developed as one takes the course. The material lends itself well to the inclusion of a course project if one desires to do so.

1

Kinematic Models for Mobile Robots

1.1 Introduction

This chapter is devoted to the development of kinematic models for two types of wheeled robots. The kinematic equations are developed along with the basic geometrical properties of achievable motion. The two configurations considered here do not exhaust the myriad of possible configurations for wheeled robots; however, they serve as an adequate test bed for the development and discussion of the principals involved.

1.2 Vehicles with Front-Wheel Steering

The first type of mobile robot to be considered is the one with front-wheel steering. Here the vehicle is usually powered via the rear wheels, and the steering is achieved by way of an actuator for turning the front wheels.

In Figure 1.1, we have a diagram for a four-wheel front-wheel-steered robot. The equations would also apply for the case of a single front wheel. The angle the front wheels make with respect to the longitudinal axis of the robot, y_{robot}, is defined as α, measured in the counter-clockwise direction. The angle that the longitudinal axis, y_{robot}, makes with respect to the y_{ground} axis is defined as ψ, also measured in the counter-clockwise direction. The instantaneous center about which the robot is turning is the point of intersection of the two lines passing through the wheel axes.

From geometry we have

$$\frac{L}{R} = \tan \alpha$$

which may be solved to yield the instantaneous radius of curvature for the path of the midpoint of the rear axle of the robot.

Mobile Robots: Navigation, Control and Sensing, Surface Robots and AUVs,
Second Edition. Gerald Cook and Feitian Zhang.
© 2020 by The Institute of Electrical and Electronics Engineers, Inc.
Published 2020 by John Wiley & Sons, Inc.

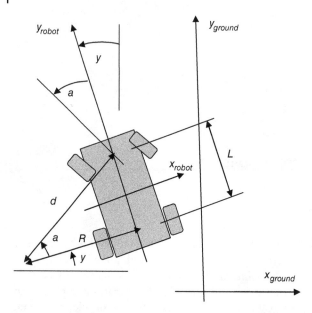

Figure 1.1 Schematic diagram of the front-wheel steered robot.

$$R = \frac{L}{\tan \alpha} \tag{1.1}$$

From geometry we also have

$$v_{rear\ wheel} = R\frac{d}{dt}(\psi) = R\dot{\psi}$$

or

$$\dot{\psi} = \frac{v_{rear\ wheel}}{R}$$

which can be written as

$$\dot{\psi} = \frac{v_{rear\ wheel}}{L/\tan\alpha} = \frac{v_{rear\ wheel}}{L}\tan\alpha \tag{1.2}$$

If one held the steering angle α constant, the trajectory would result in a circle whose radius is dictated by the robot length and the actual steering angle used per equation (1.1).

Now the instantaneous curvature itself is defined as the ratio of change in angle divided by change in distance or change in angle per distance traveled. It is given by

$$\kappa = \frac{\Delta\psi}{\Delta s} = \frac{\Delta\psi/\Delta t}{\Delta s/\Delta t} = \frac{\dot{\psi}}{v_{rear\ wheel}}$$

which is the inverse of the instantaneous radius of curvature. Thus, the radius of curvature may be interpreted as

$$R = \frac{1}{\kappa} = \frac{v_{rear\ wheel}}{\dot{\psi}} = \frac{ds}{d\psi}$$

i.e., the change in distance traveled per radian change in heading angle.

The complete set of kinematic equations for the motion in robot coordinates are

$$v_x = 0 \tag{1.3a}$$

$$v_y = v_{rear\ wheel} \tag{1.3b}$$

$$\dot{\psi} = \frac{v_{rear\ wheel}}{L} \tan \alpha \tag{1.3c}$$

Converted to earth coordinates these become

$$\dot{x} = -v_{rear\ wheel} \sin \psi \tag{1.4a}$$

$$\dot{y} = v_{rear\ wheel} \cos \psi \tag{1.4b}$$

$$\dot{\psi} = \frac{v_{rear\ wheel}}{L} \tan \alpha \tag{1.4c}$$

This form of the equations is quite simple; however, it should be noted that these equations are nonlinear. Also see Dudek and Jenkin.

Now if we wish to take into account the fact that steering angle and velocity cannot change instantaneously, we may define the derivatives or rates of these variables as control signals, i.e.,

$$\dot{\alpha} = u_1 \tag{1.5a}$$

and

$$\dot{v}_{rear\ wheel} = u_2 \tag{1.5b}$$

The system of equations for this model is now fifth order. The equations provide the correct kinematic relationships among the variables for motion and rotation in the *xy* plane but do not include the complexity of suspension or motor dynamics. Also not included in this model are robot pitch and roll.

It may be desirable to form a discrete-time model from these equations. This would be useful for discrete-time simulation as well as other applications. Clearly these equations are nonlinear. Therefore, the methods used for converting a linear continuous-time system to a discrete-time representation are not applicable. One approach is to use the Euler integration method. This method is a first-order, Taylor-series approximation to integration and says that the derivative may be approximated by a finite difference

$$\dot{x}(t) \approx \frac{x(t + \Delta t) - x(t)}{\Delta t}$$

This can be re-arranged to yield

$$x(t + \Delta t) \approx x(t) + \dot{x}(t)\Delta t$$

Setting $t = kT$ and the sampling interval $\Delta t = T$ and applying this to the above equations we have

$$x((k + 1)T) = x(kT) - Tv_{rear\ wheel}(kT)\sin\psi(kT) \tag{1.6a}$$

$$y((k + 1)T) = y(kT) + Tv_{rear\ wheel}(kT)\cos\psi(kT) \tag{1.6b}$$

$$\psi((k + 1)T) = \psi(kT) + T\frac{v_{rear\ wheel}(kT)}{L}\tan\alpha(kT) \tag{1.6c}$$

$$\alpha((k + 1)T) = \alpha(kT) + Tu_1(kT) \tag{1.6d}$$

and

$$v((k + 1)T) = v(kT) + Tu_2(kT) \tag{1.6e}$$

Here the sampling interval T must be chosen to be sufficiently small depending on the dynamics of the original differential equations, i.e., the behavior of the discrete-time model must match up with that of the original system. For a linear system, this corresponds to selecting the sampling interval to be approximately one-fifth of the smallest time constant of the system or smaller depending on the degree of precision required. For nonlinear systems, it may be necessary to determine this limiting size empirically. This discrete-time model may be used for analysis, control design, estimator design, and simulation.

It should be noted that more sophisticated and more robust methods exist for converting continuous-time dynamic system models to discrete-time models. For more information on this topic the reader is referred to *Digital Simulation of Dynamic Systems* by Hartley, Beale and Chicatelli.

From time to time, it will be convenient to interpret speed expressed in various units. For this reason the following equalities are presented.

$$10\,\text{km/h} = 2.778\,\text{m/s} = 9.1134\,\text{ft/s} = 6.2137\,\text{mph}$$

1.3 Vehicles with Differential-Drive Steering

Another common type of steering used for mobile robots is differential-drive steering illustrated in Figure 1.2. Here the wheels on one side of the robot are controlled independently of the wheels on the other side. By coordinating the two different speeds, one can cause the robot to spin in place, move in a straight line, move in a circular path, or follow any prescribed trajectory.

The equations of motion for the robot steered via differential wheel speeds are now derived. Let R represent the instantaneous radius of curvature of the robot

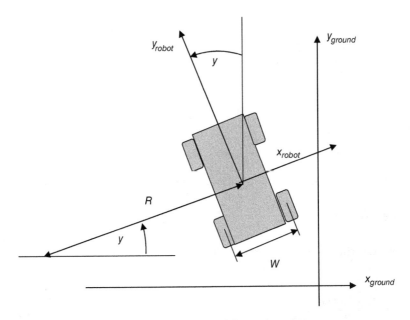

Figure 1.2 Schematic diagram of differential-drive robot.

trajectory. The width of the vehicle, i.e., spacing between the wheels, is designated as W. From geometrical considerations we have:

$$v_{left} = \dot{\psi}(R - W/2) \tag{1.7a}$$

and

$$v_{right} = \dot{\psi}(R + W/2) \tag{1.7b}$$

Now subtracting the two above equations yields

$$v_{right} - v_{left} = \dot{\psi} W$$

so we obtain for the angular rate of the robot

$$\dot{\psi} = \frac{v_{right} - v_{left}}{W} \tag{1.8}$$

Solving for the instantaneous radius of curvature, we have:

$$R = \frac{v_{left}}{\dot{\psi}} + \frac{W}{2}$$

or

$$R = \frac{v_{left}}{\dfrac{v_{right} - v_{left}}{W}} + \frac{W}{2}$$

or finally

$$R = \frac{W}{2} \frac{v_{right} + v_{left}}{v_{right} - v_{left}} \tag{1.9}$$

This results in the expression for velocity along the robot's longitudinal axis:

$$v_y = \dot{\psi}R = \frac{v_{right} - v_{left}}{W} \frac{W}{2} \frac{v_{right} + v_{left}}{v_{right} - v_{left}} = \frac{v_{right} + v_{left}}{2}$$

In summary, the equations of motion in robot coordinates are:

$$v_x = 0 \tag{1.10a}$$

$$v_y = \frac{v_{right} + v_{left}}{2} \tag{1.10b}$$

and

$$\dot{\psi} = \frac{v_{right} - v_{left}}{W} \tag{1.10c}$$

If we convert to earth coordinates these become:

$$\dot{x} = -\frac{v_{right} + v_{left}}{2} \sin\psi \tag{1.11a}$$

$$\dot{y} = \frac{v_{right} + v_{left}}{2} \cos\psi \tag{1.11b}$$

and

$$\dot{\psi} = \frac{v_{right} - v_{left}}{W} \tag{1.11c}$$

As we did in the case for the robot with front-wheel steering, we may wish to account for the fact that velocities cannot change instantaneously. Thus, we would introduce as the control variables the velocity rates:

$$\dot{v}_{right} = u_1 \tag{1.12a}$$

and

$$\dot{v}_{left} = u_2 \tag{1.12b}$$

The system of equations for this kinematic model is now fifth order.

Again we can use the Euler integration method for obtaining a discrete-time model for this system of nonlinear equations,

$$x((k+1)T) = x(kT) - T\frac{v_{right}(kT) + v_{left}(kT)}{2} \sin\psi(kT) \tag{1.13a}$$

$$y((k+1)T) = y(kT) + T\frac{v_{right}(kT) + v_{left}(kT)}{2} \cos\psi(kT) \tag{1.13b}$$

$$\psi((k+1)T) = \psi(kT) + T\frac{v_{right}(kT) - v_{left}(kT)}{W} \tag{1.13c}$$

$$v_{right}((k+1)T) = v_{right}(kT) + Tu_1(kT) \tag{1.13d}$$

and

$$v_{left}((k+1)T) = v_{left}(kT) + Tu_2(kT) \tag{1.13e}$$

More sophisticated and more accurate methods for obtaining discrete-time models exist; however, this Euler model may be quite useful if the sampling interval is set sufficiently small. These discrete-time models may be used for system analysis, controller design, estimator design, and system simulation. More complex models for mobile robots could also include pitch, roll, and vertical motion.

Exercises

1 A front-wheel steered robot is to turn to the left with a radius of curvature equal to 20 m. The robot is 1 m wide and 2 m long. What should the steering angle be?

2 A differential wheel steered robot is to turn to the left with a radius of curvature equal to 20 m and is to travel at 1 m/s. The width is 1 m and the length is 2 m. What should be the velocities of the right side and the left side?

3 Using the discrete-time model presented, perform a digital simulation of the front-wheel steered robot using a steering angle of $45°$, a length of 1.5 m, and a speed of 2.778 m/s. Experiment with the sample interval, T and find the maximum allowable value that yields consistent results.

4 Develop a digital simulation for the steered wheel robot modeled in Chapter 1. Assume that the width from wheel to wheel is 1 m and that the length, axle to axle is 2 m. A sequence of speeds and steering angles will be inputs. Include limits in your model so that steering angle will not exceed $\pm45°$ regardless of the command. Simulate the robot for straight line motion and for motion when the steering angle is held constant at $45°$ and then

constant at −45°. Simulate several seconds of motion. Use the Euler formula for integration and experiment with the sampling interval. Then use a sampling interval of 0.1 s and see if this sampling interval yields correct results. Plot x vs. t, y vs. t, heading vs. t, and y vs. x.

5 Develop a digital simulation for the differential drive robot, modeled in Chapter 1. Assume that the width from wheel to wheel is 1 m and that the length, axle to axle is 2 m. A sequence of right side speeds and left side speeds will be the inputs. Simulate for straight line motion and for motion when the right side speed is 10% above the average speed (right speed + left speed)/2 and the left side speed is 10% below the average speed. Simulate several seconds of motion. Use the Euler formula for integration and experiment with the sampling interval. Then use a sampling interval of 0.1 s and see if this sampling interval yields correct results. Plot x vs. t, y vs. t, heading vs. t, and y vs. x.

References

Canudas de Wit, Carlos, Siciliano, Bruno, and Bastin, Georges (eds.), *Theory of Robot Control*, Springer, 1996.

Corke, P. I. and Ridley, P., "Steering Kinematics for a Center-Articulated Mobile Robot," *IEEE Transactions on Robotics and Automation*, Vol. **17**, No. 2 (2001), pp. 215–218.

Dudek, Gregory and Jenkin, Michael, *Computational Principles of Mobile Robotics*, Cambridge University Press, 2000.

Fahimi, Farbod, *Autonomous Robots: Modeling, Path Planning, and Control*, Springer, 2009.

Hartley, Tom, Beale, Guy O., and Chicatelli, Stephen P., *Digital Simulation of Dynamic Systems*, Prentice Hall, 1994.

Indiveri, G., "An Introduction to Wheeled Mobile Robot Kinematics and Dynamics," *Robocamp? Padeborn (Germany)* (April 8, 2002).

Kansal, S., Jakkidi, S., and Cook, G., "The Use of Mobile Robots for Remote Sensing and Object Localization," *Proceedings of IECON 2003* (pp. 279–284, Roanoke, VA, November 2–6, 2003).

2

Mobile Robot Control

2.1 Introduction

This chapter is devoted primarily to the steering control of wheeled mobile robots, with minor attention also devoted to speed control. Different tools of control theory are applied here with attention given to various measures of performance including stability. A local coordinate system with quite general applicability is introduced. The chapter concludes with a section on minimal path length trajectories.

2.2 Front-Wheel Steered Vehicle, Heading Control

Now that mathematical models have been developed for the mobile robot, several controllers for the speed and direction of the mobile robot will be proposed and analyzed. Performance, including stability and robustness is of greatest interest. First, heading control of the front-wheel steered robot will be addressed. In the following, for simplicity of notation we set:

$$v_{rear\ wheel} = V$$

The desired heading may be given directly as a command:

$$\psi_{des} = specified\ heading$$

This direction could arise from a predetermined trajectory or it could be designated by a sensor that detected something of interest. The desired direction could also be computed in terms of the current location and the coordinates of a destination if one does not have to be concerned for obstacles. The direction from the current robot position to the destination may be expressed as:

$$\psi_{des} = -\tan^{-1}\left(\frac{x_{des} - x}{y_{des} - y}\right)$$

Mobile Robots: Navigation, Control and Sensing, Surface Robots and AUVs,
Second Edition. Gerald Cook and Feitian Zhang.
© 2020 by The Institute of Electrical and Electronics Engineers, Inc.
Published 2020 by John Wiley & Sons, Inc.

Initially it will be assumed that by some means a desired heading has been established and the goal will be to aim the robot in that direction. Several steering algorithms will now be presented.

First assuming that the steering angle may be commanded directly, we could choose as our expression for it:

$$\alpha = K(\psi_{des} - \psi) \tag{2.1}$$

which is illustrated in Figure 2.1.

The steering angle here is proportional to the error in heading, i.e., linear control. Making the approximation $\tan(\alpha) \approx \alpha$ yields for the robot closed-loop heading equation:

$$\dot{\psi} = \frac{V}{L} \tan \alpha \approx \frac{V}{L} \alpha$$

or

$$\dot{\psi} = \frac{V}{L} K(\psi_{des} - \psi) \tag{2.2}$$

Assuming a fixed velocity, the error in heading goes to zero exponentially. The speed of convergence is determined by the time constant, $T_c = L/KV$.

Another control algorithm for consideration is

$$\alpha = \frac{\pi}{4} \text{sign}(\psi_{des} - \psi) \tag{2.3}$$

which is illustrated in Figure 2.2.

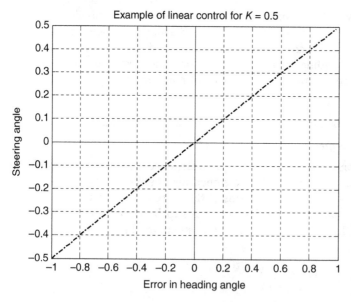

Figure 2.1 α versus $(\psi_{des} - \psi)$ for $\alpha = K(\psi_{des} - \psi)$ and $K = 0.5$.

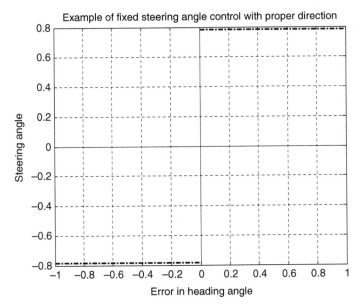

Figure 2.2 α versus $(\psi_{des} - \psi)$ for $\alpha = \dfrac{\pi}{4}\,\mathrm{sign}(\psi_{des} - \psi)$.

Here the steering angle steers in the proper direction with a fixed steering angle of 45°. Again using the approximation $\tan(\alpha) \approx \alpha$ yields

$$\dot{\psi} = \frac{V}{L}\alpha$$

or

$$\dot{\psi} = \frac{V}{L}\frac{\pi}{4}\,\mathrm{sign}(\psi_{des} - \psi) \tag{2.4}$$

The error in heading angle goes to zero as a ramp with slope $V\pi/4L$ as is shown in Figure 2.3.

Figure 2.4 shows the behavior of the steering angle versus time.

While this rapid convergence of the heading error to zero is desirable, when the error becomes very small, the slightest bit of noise or dynamic lag will cause the steering angle to switch back and forth between $\pm\pi/4$ in what is called chattering. This is an undesirable feature as it causes wear on the steering mechanism and results in inefficient longitudinal motion of the robot.

As an alternative aimed at preserving the best features of the two algorithms above, one may combine them into a single algorithm

$$\alpha = (\pi/4)\mathrm{sign}(\psi_{des} - \psi) \quad \text{whenever} \quad K|\psi_{des} - \psi| > \pi/4 \tag{2.5a}$$

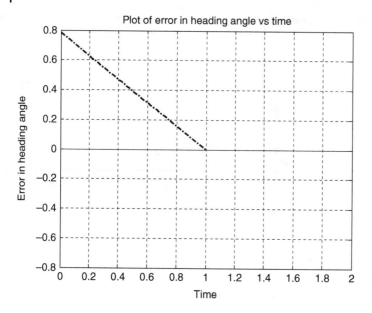

Figure 2.3 $(\psi_{des} - \psi)$ versus time for $\alpha = \dfrac{\pi}{4}\,\mathrm{sign}(\psi_{des} - \psi)$.

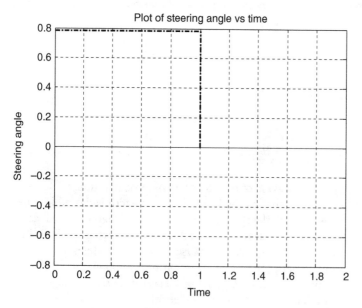

Figure 2.4 α versus time for $\alpha = \dfrac{\pi}{4}\,\mathrm{sign}(\psi_{des} - \psi)$.

and

$$\alpha = K(\psi_{des} - \psi) \quad whenever \quad K|\psi_{des} - \psi| < \frac{\pi}{4} \tag{2.5b}$$

This control algorithm is illustrated in Figure 2.5.

Inserting this steering algorithm into the linearized equation for heading rate yields

$$\dot{\psi} = \frac{V}{L}\frac{\pi}{4} \operatorname{sign}(\psi_{des} - \psi) \quad whenever \quad K|\psi_{des} - \psi| > \frac{\pi}{4} \tag{2.6a}$$

and

$$\dot{\psi} = \frac{V}{L}K(\psi_{des} - \psi) \quad whenever \quad K|\psi_{des} - \psi| < \frac{\pi}{4} \tag{2.6b}$$

The steering angle steers in the proper direction with a fixed steering angle and the error in heading initially diminishes as a ramp. Then as the gain times the error in heading becomes less than $\pi/4$ radians, the algorithm reverts to linear control with steering angle proportional to error in heading, i.e., proportional to $\psi_{des} - \psi$. The error in heading then has exponential decay and does not chatter. Figures 2.6 and 2.7 illustrate this. Note that the steering control is saturated for the first three seconds and then becomes proportional control.

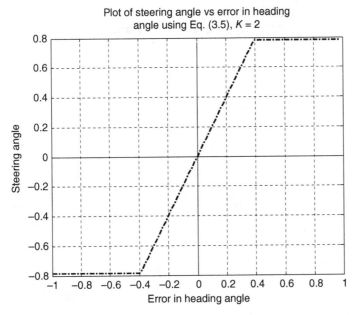

Figure 2.5 α versus $(\psi_{des} - \psi)$ for the control algorithm of equation (2.5) with $K = 2$.

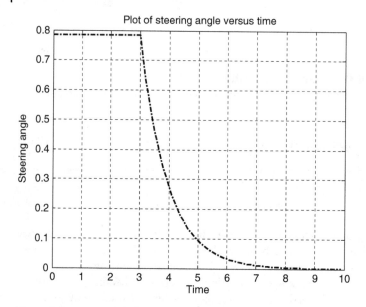

Figure 2.6 α versus time for $\psi(0) = 0$ and $\psi_{des} = \pi/2$, $K = 2.0$.

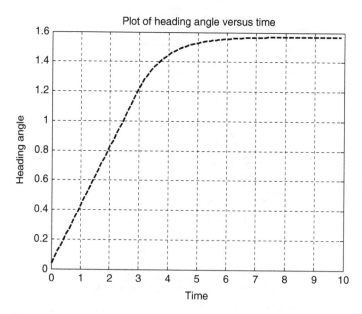

Figure 2.7 ψ versus time for $\psi(0) = 0$ and $\psi_{des} = \pi/2$, $K = 2.0$.

Note also that the heading angle is a ramp for the first three seconds and then approaches the desired heading with exponential error decay. Here $V = 1$ and $L = 2$.

Care must be used when basing the control on an angle to ensure that the control action for a given angle is consistent even if the expression for the angle exceeds 2π in magnitude. For the above steering angle algorithm one may create a variable, *angle. error* $= \psi_{des} - \psi$. If *angle. error* $> 2\pi$, set *angle. error* $=$ *angle. error* $- 2\pi$. On the other hand, if *angle. error* $< -2\pi$, set *angle. error* $=$ *angle. error* $+ 2\pi$. The control action is then based on this new expression for *angle. error*.

Example 1 *Demonstrate the steering strategy just described for the case of driving the mobile robot to a destination in the x–y space. Let $\psi(0) = -\dfrac{\pi}{2}$, $x(0) = 0$, and $y(0) = 0$. The destination is $x_{dest} = 10$ and $y_{dest} = 10$. Use a gain of $K = 2.0$.*

Solution 1

Here the desired heading varies as it is computed based on the current location of the mobile robot and the destination, i.e., $\psi_{des}(t) = -tan^{-1}[(x_{dest} - x(t))/(y_{dest} - y(t))]$. The steering control described above is then implemented and the motion of the robot simulated. The results are shown in Figure 2.8a–c.

This steering strategy works fine as long as all disturbances occur before the mobile robot gets within a distance of $L/\tan \alpha_{max}$ from the destination. Disturbances after that cannot be accommodated because of the finite turning radius of the robot.

We now assume that the steering angle alpha may not be commanded directly but rather that its rate is the controlled variable. Setting the rate of the steering angle proportional to error in heading yields

$$\dot{\alpha} = K(\psi_{des} - \psi) \tag{2.7}$$

which again is linear control. Taking the equation for heading rate under the assumption of small steering angle and differentiating yields

$$\ddot{\psi} = \frac{V}{L}\dot{\alpha} \tag{2.8}$$

or

$$\ddot{\psi} = K\frac{V}{L}(\psi_{des} - \psi) \tag{2.9}$$

Unfortunately this control algorithm yields imaginary poles leading to sustained oscillations. This may be seen by applying the Laplace Transform to the above equation yielding

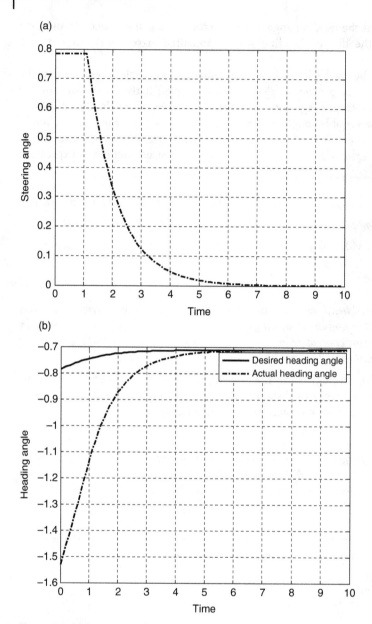

Figure 2.8 (a) Steering angle α versus time. (b) Desired heading and actual heading versus time. (c) Resulting trajectory, y versus x.

(c)

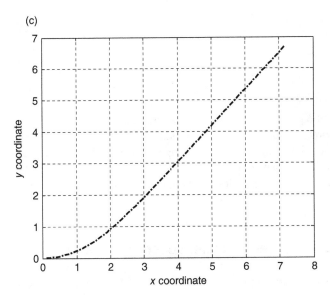

Figure 2.8 (Continued)

$$\psi(s) = \frac{KV/L}{s^2 + KV/L}\psi_{des}(s)$$

Introducing an additional term into the algorithm, this time a feedback term in the steering angle itself, we consider as a steering algorithm

$$\dot{\alpha} = K_1(\psi_{des} - \psi) - K_2\frac{V}{L}\alpha \qquad (2.10)$$

The steering angle rate is proportional to heading error and steering angle. Using the approximate expression for heading rate,

$$\dot{\psi} = \frac{V}{L}\alpha$$

the above equation for the steering angle rate may be re-expressed as

$$\dot{\alpha} = K_1(\psi_{des} - \psi) - K_2\dot{\psi} \qquad (2.11)$$

The final expression for the steering algorithm, equation (2.11), reveals that in fact the additional term in the control algorithm adds damping through rate feedback. For implementation purposes it is preferable to use the former expression, equation (2.10), which is in terms of steering angle since the latter expression would require differentiation of heading angle.

Using this control algorithm which contains damping in the equation for $\ddot{\psi}$ above yields

$$\ddot{\psi} = K_1 \frac{V}{L}(\psi_{des} - \psi) - K_2 \frac{V}{L}\dot{\psi} \qquad (2.12)$$

which corresponds to

$$\psi(s) = \frac{(K_1 V/L)}{s^2 + (K_2 V/L)s + (K_1 V/L)}\psi_{des}(s)$$

By proper choice of K_1 and K_2, one can achieve any desired pole locations. In implementing the above, one must use care to ensure that neither the maximum steering angle rate nor the maximum steering angle is exceeded. Also it should be kept in mind that the performance analysis has been done under the assumption of sufficiently small steering angles allowing one to make the approximation, $\tan(\alpha) \approx \alpha$.

2.3   Front-Wheel Steered Vehicle, Speed Control

So far the focus has been on controlling the heading of the mobile robot for steering purposes. In addition, one must control the robot speed. One way to select the desired speed for the robot would be in terms of distance to the destination and time remaining, i.e.,

$$V_{des} = \frac{\sqrt{(x_{des} - x)^2 + (y_{des} - y)^2}}{time_{to\ go}}$$

Now it is possible that the desired velocity would exceed the velocity achievable by the robot. In that case, one could modify the algorithm for commanded velocity to become

$$V_{des} = \min\left\{\frac{\sqrt{(x_{des} - x)^2 + (y_{des} - y)^2}}{time_{to\ go}}, V_{max}\right\} \qquad (2.13)$$

Here velocity is commanded to be (distance remaining)/(time-to-go) or V_{max}, whichever is less, i.e., the control strategy is saturating command.

The above expressions assumed that velocity could jump to the commanded value instantaneously. A more realistic approach would be to command a rate to achieve the desired velocity or to have the velocity approach the desired velocity with a time constant. The expression for the latter method

$$\tau\dot{V} + V = V_{des} \qquad (2.14)$$

can be re-arranged to

$$\dot{V} = \frac{V_{des} - V}{\tau}$$

or

$$\dot{V} = \frac{\min\left\{ \dfrac{\sqrt{(x_{des} - x)^2 + (y_{des} - y)^2}}{time_{to\ go}}, V_{max} \right\} - V}{\tau} \tag{2.15}$$

Here desired velocity has been expressed as the saturating command, (distance remaining)/(time-to-go) or V_{max}, whichever is less. The velocity approaches the desired value with a time constant of τ.

In all of these speed control possibilities, it may be of interest to also compute the energy consumed. If one can determine the motor torque required to achieve the motion, then since torque is proportional to armature current, one would be able to compute the electrical power consumed (current times voltage) and integrate this to find energy consumed. Energy management is especially important when operating from a finite energy supply such as that of a battery.

To integrate turn and velocity control for the front-wheel steered robot, one can simply have the robot turn as it travels longitudinally, and the somewhat separate controls jointly affect the vehicle trajectory. The actual path mapped out on the ground for a fixed steering angle is independent of the robot speed. Therefore, for many problems it is reasonable to separate the two control problems. If there are obstacles or if high speeds could cause the robot to skid during steep turns, then steering and velocity controls must be coordinated.

2.4 Heading and Speed Control for the Differential-Drive Robot

The robot with differential wheel control turns by using different wheel speeds. One approach to steering toward a particular heading would be to rotate in place making use of the equation for heading rate

$$\dot{\psi} = \frac{v_{right} - v_{left}}{W}$$

and picking

$$v_{right} = V_{max}$$

and

$$v_{left} = -V_{max}$$

until achieving the desired heading. Then the robot could proceed forward toward the destination at the desired speed making use of the fact that

$$V = \frac{v_{right} + v_{left}}{2}$$

and using

$$v_{right} = V_{desired}$$

and also

$$v_{left} = V_{desired}$$

For a commanded turn in place at a rate that is lower than the maximum turn rate, one may determine the appropriate wheel velocities using

$$\dot{\psi}_{desired} = \frac{v_{right\ desired} - v_{left\ desired}}{W} \tag{2.16}$$

and solving to obtain

$$v_{right} = \frac{W \dot{\psi}_{desired}}{2} \tag{2.17a}$$

and

$$v_{left} = -\frac{W \dot{\psi}_{desired}}{2} \tag{2.17b}$$

This combination achieves the desired turn rate and also causes the longitudinal velocity to be zero during the turn. After achieving the desired heading, one may use an algorithm of the type used before

$$v_{right} = \min \left\{ \frac{\sqrt{(x_{des} - x)^2 + (y_{des} - y)^2}}{time_{to\ go}}, V_{max} \right\} \tag{2.18a}$$

and

$$v_{left} = \min \left\{ \frac{\sqrt{(x_{des} - x)^2 + (y_{des} - y)^2}}{time_{to\ go}}, V_{max} \right\} \tag{2.18b}$$

These wheel velocities yield the correct robot velocity and also maintain straight-line motion. Upon reaching the destination, one could then rotate in place to obtain the desired final heading. If one wishes to combine turning of the robot with longitudinal motion, then some strategy must be first used to determine the desired turn rate and the desired velocity. Once these have been obtained, the equations for turn rate and longitudinal velocity can be combined to yield as solutions

$$v_{right} = V_{desired} + \frac{W \dot{\psi}_{desired}}{2} \tag{2.19a}$$

and

$$v_{left} = V_{desired} - \frac{W\dot{\psi}_{desired}}{2}$$
(2.19b)

It is easy to show that this combination satisfies both the desired turn rate and desired longitudinal velocity.

The following compares two contrasting strategies for reaching a destination. One is to spread the rotation uniformly over the entire trajectory and turn while moving at the maximum speed (*turn-while-traveling*). The other is to turn in place at maximum rotational speed and then travel at maximum speed to the destination (*turn-then-travel*). Let the initial heading offset be given by $\Delta\psi/2$ and the distance between the initial and final points be given by D. This is shown in the construction presented in Figure 2.9.

For the *turn-while-traveling* strategy the path is a circle. From the law of sines one can determine the relationship between R and D to be

$$R = D/(2\sin(\Delta\psi/2))$$

The travel time can be found by realizing that the outer wheels travel at speed V_{max} and travel a distance $(R + W/2)\Delta\psi$. Thus the travel time is

$$T_{turn\ while\ traveling} = \frac{(R + W/2)\Delta\psi}{V_{max}} = \frac{W\Delta\psi}{2V_{max}} + \frac{R\Delta\psi}{V_{max}}$$

Note that the final heading of the mobile robot is $-\Delta\psi/2$.

For the *turn-then-travel* strategy, including turning to the same heading at the end of the trajectory as the *turn-while-traveling* strategy, the total time required is

$$T_{turn\ then\ travel} = \frac{W\Delta\psi/2}{2V_{max}} + \frac{D}{V_{max}} + \frac{W\Delta\psi/2}{2V_{max}} = \frac{W\Delta\psi}{2V_{max}} + \frac{2R\sin(\Delta\psi/2)}{V_{max}}$$

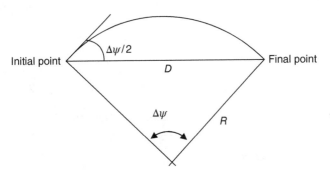

Figure 2.9 Construction for evaluating travel time.

The first terms in each expression are identical, but the last term is smaller for the case of the *turn-then-travel* strategy. The greater $\Delta\psi/2$ is, the greater this difference will be.

2.5 Reference Trajectory and Incremental Control, Front-Wheel Steered Robot

At this point, we shall examine the mobile robot control problem using a method called incremental control about a reference trajectory. Rather than robot heading control as was just addressed, here the goal will be to steer the robot so as to cause it to move along a specified reference trajectory. The equations of motion for the front-wheel steered robot with rear-wheel drive are repeated below for convenience.

$$\dot{x} = -V \sin \psi$$
$$\dot{y} = V \cos \psi$$

and

$$\dot{\psi} = \frac{V}{L} \tan \alpha$$

Here we are using the simplified third-order model which assumes that the velocity and steering angle can be directly controlled. As was already noted, the dynamic equations are seen to be nonlinear, whereas most of the theory for control design is based on linear systems. One approach to controlling a nonlinear system is to first define a reference trajectory satisfying the overall objectives of the problem at hand. This trajectory comprises values of the system state and control at all points between the initial and final conditions. These values for the reference trajectory will be denoted with the subscript r.

$$\dot{x}_r = -V_r \sin \psi_r \tag{2.20a}$$
$$\dot{y}_r = V_r \cos \psi_r \tag{2.20b}$$

and

$$\dot{\psi}_r = \frac{V_r}{L} \tan \alpha_r \tag{2.20c}$$

Next we subtract the equations describing the reference trajectory from the original equations.

$$\dot{x} - \dot{x}_r = -V \sin \psi + V_r \sin \psi_r \tag{2.21a}$$
$$\dot{y} - \dot{y}_r = V \cos \psi - V_r \cos \psi_r \tag{2.21b}$$

and

$$\dot{\psi} - \dot{\psi}_r = \frac{V}{L} \tan \alpha - \frac{V_r}{L} \tan \alpha_r \tag{2.21c}$$

Next we expand each term in the equations above in a Taylor Series about the reference values and drop all the terms of second order and higher. Defining the deviations from the reference values as δ, we obtain an incremental model

$$\delta \dot{x} = -V \sin \psi + V_r \sin \psi_r \approx -V_r \sin \psi_r - \delta V \sin \psi_r$$
$$- V_r \cos \psi_r \delta \psi + V_r \sin \psi_r$$

or

$$\delta \dot{x} \approx -\delta V \sin \psi_r - V_r \cos \psi_r \delta \psi \tag{2.22a}$$

$$\delta \dot{y} = V \cos \psi - V_r \cos \psi_r \approx V_r \cos \psi_r - V_r \sin \psi_r \delta \psi + \delta V \cos \psi_r$$
$$- V_r \cos \psi_r$$

or

$$\delta \dot{y} \approx \delta V \cos \psi_r - V_r \sin \psi_r \delta \psi \tag{2.22b}$$

and

$$\delta \dot{\psi} = \frac{V}{L} \tan \alpha - \frac{V_r}{L} \tan \alpha_r \approx \frac{V_r}{L} \tan \alpha_r + \frac{V_r}{L \cos^2 \alpha} \delta \alpha + \frac{\delta V}{L} \tan \alpha_r - \frac{V_r}{L} \tan \alpha_r$$

or

$$\delta \dot{\psi} \approx \frac{V_r}{L \cos^2 \alpha} \delta \alpha + \frac{\delta V}{L} \tan \alpha_r \tag{2.22c}$$

Having retained only the first-order terms, these can be re-arranged as

$$\begin{bmatrix} \delta \dot{x} \\ \delta \dot{y} \\ \delta \dot{\psi} \end{bmatrix} = \begin{bmatrix} 0 & 0 & -V_r \cos \psi_r \\ 0 & 0 & -V_r \sin \psi_r \\ 0 & 0 & 0 \end{bmatrix} \begin{bmatrix} \delta x \\ \delta y \\ \delta \psi \end{bmatrix} + \begin{bmatrix} -\sin \psi_r & 0 \\ \cos \psi_r & 0 \\ \dfrac{\tan \alpha_r}{L} & \dfrac{V_r}{L \cos^2 \alpha_r} \end{bmatrix} \begin{bmatrix} \delta V \\ \delta \alpha \end{bmatrix}$$

$$\tag{2.23}$$

The terms within the coefficient matrices in the above equations would be evaluated along the reference trajectory. This incremental model is equivalent to a time-varying linear system.

$$\begin{bmatrix} \delta \dot{x} \\ \delta \dot{y} \\ \delta \dot{\psi} \end{bmatrix} = [A(t)] \begin{bmatrix} \delta x \\ \delta y \\ \delta \psi \end{bmatrix} + [B(t)] \begin{bmatrix} \delta V \\ \delta \alpha \end{bmatrix}$$

If the coefficient matrices remain constant, then it becomes a time-invariant linear system.

$$
\begin{bmatrix} \delta \dot{x} \\ \delta \dot{y} \\ \delta \dot{\psi} \end{bmatrix} = [A] \begin{bmatrix} \delta x \\ \delta y \\ \delta \psi \end{bmatrix} + [B] \begin{bmatrix} \delta V \\ \delta \alpha \end{bmatrix}
$$

In this case one can utilize the theory of linear control to design a controller to maintain the mobile robot near the reference trajectory. Note that for all coefficients of A and B to be constant, V_r, ψ_r, and α_r must be constant implying that α_r is zero. A very simple linear control algorithm for the above would be

$$
\delta V = -K_1 \delta y
$$

and

$$
\delta \alpha = K_2 \delta x - K_3 \delta \psi
$$

Here the positive dependence of $\delta \alpha$ on δx is selected because the x axis is positive to the right and positive α steers to the left.

The above equations then become in closed loop

$$
\begin{bmatrix} \delta \dot{x} \\ \delta \dot{y} \\ \delta \dot{\psi} \end{bmatrix} = \begin{bmatrix} 0 & K_1 \sin \psi_r & -V_r \cos \psi_r \\ 0 & -K_1 \cos \psi_r & -V_r \sin \psi_r \\ K_2 \dfrac{V_r}{L} & 0 & -K_3 \dfrac{V_r}{L} \end{bmatrix} \begin{bmatrix} \delta x \\ \delta y \\ \delta \psi \end{bmatrix} \tag{2.24}
$$

One may now determine the eigenvalue equation for the above matrix

$$
\lambda^3 + (K_1 \cos \psi_r + K_3 V_r/L)\lambda^2 + (K_2 V_r^2 \cos \psi_r/L + K_1 K_3 V_r \cos \psi_r/L)\lambda
$$
$$
+ (K_1 K_2 V_r^2/L) = 0
$$

and compare it to that corresponding to the desired eigenvalues

$$
(\lambda - \lambda_{1des})(\lambda - \lambda_{2des})(\lambda - \lambda_{3des}) = 0
$$

or

$$
\lambda^3 + (-\lambda_{1des} - \lambda_{2des} - \lambda_{3des})\lambda^2 + (\lambda_{1des}\lambda_{2des} + \lambda_{1des}\lambda_{3des} + \lambda_{2des}\lambda_{3des})
$$
$$
\lambda - \lambda_{1des}\lambda_{2des}\lambda_{3des} = 0
$$

By equating the coefficients of these two equations, one can solve for the K_i's. It should be clear that by the proper choice of the K_i's one can achieve any desirable eigenvalues. The equations for the incremental model simplify considerably for the case where the trajectory is a straight line along the y axis. Here both $\psi_r = 0$ and $\alpha_r = 0$ causing the equations to become

$$
\begin{bmatrix} \delta \dot{x} \\ \delta \dot{y} \\ \delta \dot{\psi} \end{bmatrix} = \begin{bmatrix} 0 & 0 & -V_r \\ 0 & 0 & 0 \\ 0 & 0 & 0 \end{bmatrix} \begin{bmatrix} \delta x \\ \delta y \\ \delta \psi \end{bmatrix} + \begin{bmatrix} 0 & 0 \\ 1 & 0 \\ 0 & \dfrac{V_r}{L} \end{bmatrix} \begin{bmatrix} \delta V \\ \delta \alpha \end{bmatrix} \tag{2.25}
$$

It should be noted that for any straight-line reference path, one could translate and rotate the axes so that the reference path would be along the y axes and the assumptions of $\psi_r = 0$ and $\alpha_r = 0$ would be satisfied. We use the same simple linear control algorithm as was used above

$$
\delta V = -K_1 \delta y \tag{2.26a}
$$

and

$$
\delta \alpha = K_2 \delta x - K_3 \delta \psi \tag{2.26b}
$$

The coefficient matrix for the closed-loop incremental system becomes

$$
\begin{bmatrix} \delta \dot{x} \\ \delta \dot{y} \\ \delta \dot{\psi} \end{bmatrix} = \begin{bmatrix} 0 & 0 & -V_r \\ 0 & -K_1 & 0 \\ \dfrac{K_2 V_r}{L} & 0 & -\dfrac{K_3 V_r}{L} \end{bmatrix} \begin{bmatrix} \delta x \\ \delta y \\ \delta \psi \end{bmatrix} \tag{2.27}
$$

The associated eigenvalue equation for the closed-loop coefficient matrix becomes

$$
(\lambda + K_1)\left(\lambda^2 + \frac{K_3 V_r}{L} \lambda + \frac{K_2 V_r^2}{L} \right) = 0
$$

It is easy to show that the solutions will all lie in the left-half plane as long as all the gains, K_1, K_2, and K_3 are positive and that one can choose these coefficients to place the eigenvalues wherever one pleases.

Alternatively, one can analyze the stability of this control algorithm by utilizing Lyapunov Stability Theory. The linearized equations under the stated conditions for the simplified case are repeated for convenience

$$
\delta \dot{x} \approx -V_r \delta \psi
$$
$$
\delta \dot{y} \approx \delta V
$$

and

$$
\delta \dot{\psi} \approx \frac{V_r}{L} \delta \alpha
$$

Now using the control algorithm previously introduced,

$$
\delta V = -K_1 \delta y
$$

and

$$\delta\alpha = K_2\delta x - K_3\delta\psi$$

we have for the closed–loop system equations

$$\delta\dot{x} \approx -V_r\delta\psi \qquad (2.28a)$$
$$\delta\dot{y} \approx -K_1\delta y \qquad (2.28b)$$

and

$$\delta\dot{\psi} \approx \frac{K_2V_r}{L}\delta x - \frac{K_3V_r}{L}\delta\psi \qquad (2.28c)$$

Now taking as a Lyapunov function

$$LF = \frac{1}{2}\left[a\delta x^2 + b\delta y^2 + c\delta\psi^2\right] \qquad (2.29)$$

with a, b, and c all being positive, we have

$$d(LF)/dt = a\delta x(-V_r\delta\psi) + b\delta y(-K_1\delta y) + c\delta\psi\left(\frac{K_2V_r}{L}\delta x - \frac{K_3V_r}{L}\delta\psi\right)$$

or

$$d(LF)/dt = -aV_r\delta x\delta\psi - bK_1\delta y^2 + \frac{cK_2V_r}{L}\delta x\delta\psi - c\frac{K_3V_r}{L}\delta\psi^2$$

or

$$d(LF)/dt = -bK_1\delta y^2 + \left(\frac{cK_2}{L} - a\right)V_r\delta x\delta\psi - c\frac{K_3V_r}{L}\delta\psi^2 \qquad (2.30)$$

By taking

$$a = \frac{cK_2}{L}$$

which implies that K_2 is positive, this reduces to

$$d(LF)/dt = -bK_1\delta y^2 - c\frac{K_3V_r}{L}\delta\psi^2 \qquad (2.31)$$

which is negative semidefinite for positive values for b, c, K_1, and K_3. The only nonzero states for which $d(LF)/dt$ is zero would be when δx is nonzero and both δy and $\delta\psi$ are zero. The control action guarantees that for the closed loop system this is not an equilibrium state. Thus the system is proven to be asymptotically stable within the accuracy of the linearized model.

2.6 Heading Control of Front-Wheel Steered Robot Using the Nonlinear Model

It is of interest to study the stability of steering control laws without using any linear approximations of the equations of motion. Here we will again assume that one may manipulate the steering angle α directly. The desired heading is taken to be the direction from the current robot position to a desired fixed location. It is measured with respect to the y axis in the counter-clockwise direction. In addition to retaining the nonlinearities of the robot model, we also incorporate the fact that the desired heading angle changes with robot motion. The expression for the desired heading is

$$\psi_{des} = \tan^{-1}\left[\frac{-(x_{des}-x)}{(y_{des}-y)}\right]$$

Here it is seen that the x term enters as a negative since x is defined positive to the right and ψ is defined positive in the counter-clockwise direction or to the left. Note that the coordinates may be shifted so that the desired location is at the origin of x–y space. Define

$$X = x - x_{des} \tag{2.32a}$$

and

$$Y = y - y_{des} \tag{2.32b}$$

Noting that for a fixed desired location

$$\dot{x}_{des} = 0$$

and

$$\dot{y}_{des} = 0$$

we obtain in the new coordinates the same model as before

$$\dot{X} = -V \sin \psi$$
$$\dot{Y} = V \cos \psi$$

and

$$\dot{\psi} = \frac{V}{L} \tan \alpha$$

where the desired location in these coordinates is now the origin. The expression for the desired heading angle as given above reduces to

$$\psi_{des} = \tan^{-1}\left[\frac{X}{-Y}\right] \tag{2.33}$$

We now choose as a Lyapunov-like function

$$LF = \frac{1}{2}(\psi - \psi_{des})^2 \tag{2.34}$$

We say "Lyapunov-like" because the function is not positive definite but rather positive semidefinite. We now seek to determine a control to guarantee that

$$\frac{d}{dt}LF < 0$$

Thus, we examine

$$d/dt\left\{\frac{1}{2}(\psi - \psi_{des})^2\right\} = (\psi - \psi_{des})(d/dt\psi - d/dt\psi_{des})$$

which is a measure of the rate at which the robot converges to the desired heading. Since any error in heading causes $\frac{\pi}{2}(\psi - \psi_{des})^2$ to be positive, a large negative value for its derivative would indicate fast convergence to the desired heading. Noting that

$$d\psi_{des}/dt = \cos^2\psi_{des}\left[\frac{-Y\dot{X} + X\dot{Y}}{Y^2}\right]$$

a series of manipulations results in

$$d\psi_{des}/dt = \frac{1}{\sqrt{X^2 + Y^2}}(-V\cos\psi_{des}\sin\psi + V\sin\psi_{des}\cos\psi)$$

or

$$d\psi_{des}/dt = \frac{-V}{\sqrt{X^2 + Y^2}}\sin(\psi - \psi_{des}) \tag{2.35}$$

Using this result yields

$$d/dt\left\{\frac{1}{2}(\psi - \psi_{des})^2\right\} = (\psi - \psi_{des})\left(d\psi/dt + \frac{V}{\sqrt{X^2 + Y^2}}\sin(\psi - \psi_{des})\right)$$

or

$$d/dt\left\{\frac{1}{2}(\psi - \psi_{des})^2\right\} = \frac{V}{\sqrt{X^2 + Y^2}}(\psi - \psi_{des})\sin(\psi - \psi_{des}) + \frac{V}{L}(\psi - \psi_{des})\tan\alpha \tag{2.36}$$

It is clear that the first term is positive for heading errors less than π in magnitude. Thus, for stability the second term must be made negative and larger in magnitude than the first. One solution is to select

$$\alpha = \frac{\pi}{4}\operatorname{sign}(\psi_{des} - \psi) \tag{2.37}$$

Since $\tan(\pi/4) = 1$ and

$$(\psi_{des} - \psi)\operatorname{sign}(\psi_{des} - \psi) = |\psi_{des} - \psi|$$

this results in

$$d/dt\left\{\frac{1}{2}(\psi - \psi_{des})^2\right\} = \frac{V}{\sqrt{X^2 + Y^2}}(\psi - \psi_{des})\sin(\psi - \psi_{des}) - \frac{V}{L}|\psi - \psi_{des}| \tag{2.38}$$

which is clearly negative as long as the robot is at a distance of more than L from the destination. Thus, disturbances that occur before the robot gets within L from the final destination can be accommodated. Those that occur later cannot be accommodated. The controller here is a bang–bang controller, which does a good job of getting the robot headed in the proper direction quickly; however, as mentioned earlier, it could cause chattering, or rapid switching of the control, after reaching the desired heading. Another solution is to take

$$\tan \alpha = \sin(\psi_{des} - \psi) \tag{2.39}$$

This strategy results in

$$d/dt\left\{\frac{1}{2}(\psi - \psi_{des})^2\right\} = \frac{V}{\sqrt{X^2 + Y^2}}(\psi - \psi_{des})\sin(\psi - \psi_{des})$$
$$- \frac{V}{L}(\psi - \psi_{des})\sin(\psi - \psi_{des}) \tag{2.40}$$

which also is negative as long as the robot is at a distance of more than L from the destination. Note that since

$$\lfloor \sin(\psi_{des} - \psi)\rfloor \leq 1$$

one will have

$$|\alpha| \leq \frac{\pi}{4}$$

which is a very reasonable range for allowable steering angles. This controller outputs a smaller control signal as the heading error becomes smaller. It thus would not chatter as the previous controller did, but neither does it converge to the correct heading as quickly. Another solution would be to combine these algorithms and use

$$\alpha = \frac{\pi}{4}\operatorname{sign}(\psi_{des} - \psi) \tag{2.41a}$$

whenever the heading error exceeds some given threshold and to use

$$\tan \alpha = \sin (\psi_{des} - \psi) \tag{2.41b}$$

whenever the heading error is equal to or less than the threshold. This combines the benefits of the two previous controllers. It provides rapid convergence toward the proper heading when the heading error is large and also eliminates the chattering that would result from application of the first algorithm alone.

These solutions steer the robot toward the destination. Disturbances that occur at distances greater than L from the destination can be accommodated. The final orientation of the robot has been left free for this problem with the objective being simply to keep the robot headed toward the destination. Final orientation will depend on the initial conditions as well as the disturbances that occur during the transition.

2.7 Computed Control for Heading and Velocity, Front-Wheel Steered Robot

The model for the front-wheel steered robot is repeated for convenience.

$$\dot{X} = -V \sin \psi$$
$$\dot{Y} = V \cos \psi$$

and

$$\dot{\psi} = \frac{V}{L} \tan \alpha$$

Let us assume for this analysis that the steering angle and velocity cannot be changed instantaneously and therefore we take as the control variables

$$\dot{\alpha} = u_1$$

and

$$\dot{V} = u_2$$

i.e., we use the fifth-order model for the robot. Here the steering system may be made to behave as a second-order system with specified natural frequency and damping ratio.

For this purpose, u_1 is selected so as to achieve

$$\psi'' + 2\xi\omega_n\psi' + \omega_n^2\psi = \omega_n^2\psi_{des} \tag{2.42a}$$

or

$$\psi'' = \omega_n^2(\psi_{des} - \psi) - 2\zeta\omega_n\psi' \tag{2.42b}$$

where the prime denotes differentiation with respect to distance traveled, s. The dynamic behavior of heading is specified in this way rather than in terms of time because the actual turning motion for the front-wheel steered robot is in fact dependent on distance traveled. Noting that

$$\psi' = \frac{d\psi/dt}{ds/dt} = \frac{\tan \alpha}{L}$$

and that

$$\psi'' = \frac{1}{LV \cos^2 \alpha} \dot{\alpha}$$

the solution becomes

$$\frac{\dot{\alpha}}{LV \cos^2 \alpha} = \left[-2\zeta\omega_n \tan \alpha/L + \omega_n^2(\psi_{des} - \psi) \right]$$

or

$$u_1 = LV \cos^2 \alpha \left[-2\zeta\omega_n \tan \alpha/L + \omega_n^2(\psi_{des} - \psi) \right] \tag{2.43}$$

Note that here the system has not been approximated as a linear system, but rather the nonlinearities have been retained in the model. This type of controller is sometimes referred to as "computed control." It cancels out the existing dynamics and replaces it with the desired dynamics. It assumes that one has a perfect model of the robot. This equation for the control must be subjected to the respective constraints on maximum steering angle rate and maximum steering angle, i.e.,

$$|u_1| \leq u_{1\,max}$$

and

$$|\alpha| \leq \alpha_{max}$$

For speed control u_2 is selected so that the speed converges to the desired speed according to the solution of the first-order differential equation,

$$\tau \dot{V} + V = V_{des} \tag{2.44a}$$

or

$$u_2 = \frac{V_{des} - V}{\tau} \tag{2.44b}$$

If the values for ψ_{des} and V_{des} are constant, then the control algorithms just described guarantees a stable system as long as saturation does not occur. When in the teleoperated mode, the expression for ψ_{des} may be an input from the operator, possibly by use of a joystick. The speed command may be also be an input from the operator. Further analysis would be required in the case of time varying ψ_{des} and V_{des} or if there is saturation of the control variables.

2.8 Heading Control of Differential-Drive Robot Using the Nonlinear Model

It is of interest to analyze the stability of steering control laws for the differential-steered robot also without linearizing the equations of motion. Here we will assume that one may directly manipulate the right and left velocities. The analysis proceeds exactly as in Section 2.5 except that one replaces the equation for heading angle rate and the equation for velocity with those for the differential-drive model, i.e.,

$$\frac{V}{L} \tan \alpha \leftarrow \frac{v_{right} - v_{left}}{W}$$

and

$$V \leftarrow \frac{v_{right} + v_{left}}{2}$$

Proceeding in a parallel fashion it can be shown that one stable solution with favorable properties would be to use

$$\frac{v_{right} - v_{left}}{W} = \text{sign}(\psi_{des} - \psi) \tag{2.45a}$$

and

$$\frac{v_{right} + v_{left}}{2} = V_{des} \tag{2.45b}$$

whenever the heading error exceeds some given threshold and to use

$$\frac{v_{right} - v_{left}}{W} = \sin(\psi_{des} - \psi) \tag{2.46a}$$

and

$$\frac{v_{right} + v_{left}}{2} = V_{des} \tag{2.46b}$$

whenever the heading error is equal to or less than the threshold. The solutions to the two equations would yield the required velocities for each side. This control strategy does not induce chattering and yet provides rapid convergence when heading error is large. Simultaneously it provides the specified velocity.

2.9 Computed Control for Heading and Velocity, Differential-Drive Robot

For convenience, we repeat here the model for the differential-drive robot for the case where we take the rate change of wheel velocities as the control inputs.

$$\dot{x} = -\frac{v_{right} + v_{left}}{2} \sin \psi$$

$$\dot{y} = \frac{v_{right} + v_{left}}{2} \cos \psi$$

$$\dot{\psi} = \frac{v_{right} - v_{left}}{W}$$

$$\dot{v}_{right} = u_1$$

and

$$\dot{v}_{left} = u_2$$

The steering system here may also be made to behave as a second-order system with specified natural frequency and damping ratio. Here u_1 and u_2 are selected so as to achieve

$$\ddot{\psi} + 2\zeta\omega_n\dot{\psi} + \omega_n^2\psi = \omega_n^2\psi_{des}$$

and

$$\tau\dot{V} + V = V_{des}$$

Proceeding in a parallel fashion as in the case of the front-wheel steered robot, we arrive at the conclusions that

$$\frac{u_1 - u_2}{W} = \left[-2\zeta\omega_n\dot{\psi} + \omega_n^2(\psi_{des} - \psi) \right] \tag{2.47a}$$

and

$$\frac{u_1 + u_2}{2} = \frac{1}{\tau}\left(V_{des} - \left(v_{right} + v_{left}\right)/2\right) \tag{2.47b}$$

or

$$u_1 = \frac{V_{des} - \left(v_{right} + v_{left}\right)/2}{\tau} - \zeta\omega_n\left(v_{right} - v_{left}\right) + \omega_n^2(\psi_{des} - \psi)\frac{W}{2} \tag{2.48a}$$

and

$$u_2 = \frac{v_{des} - \left(v_{right} + v_{left}\right)/2}{\tau} + \zeta\omega_n\left(v_{right} - v_{left}\right) - \omega_n^2(\psi_{des} - \psi)\frac{W}{2} \tag{2.48b}$$

This illustrates the application of computed control to the differential-drive robot. It assumes a perfect model for the robot, and provides a control to cancel the nonlinearities. It then further includes the necessary terms to give the specified behavior.

2.10 Steering Control Along a Path Using a Local Coordinate Frame

The objective of the approach to be discussed here is tracking along a curved path without prior knowledge of the path. This is in contrast with the reference trajectory approach where prior knowledge of the path is presumed and the equations were linearized. It also differs from the situation where one heads toward a particular destination and is unconstrained by the need to stay on a path. The local coordinate frame to be used here is defined such that the x axis of the local frame passes through the robot and is normal to the path. The origin is on the path with coordinates designated as x_{path} and y_{path}, and the y axis is tangent to the path at the origin and is pointed in the direction of the desired travel. See Figure 2.10. As the vehicle moves along near the path, the origin of the coordinate system moves along the path maintaining the relationship described between the robot position and the position and orientation of the local coordinates.

The coordinates of the path itself are given as functions of the distance which the origin of the local coordinate system has traveled along the path as the robot moves, i.e.,

$$x_{path} = x_{path}\left(s_{path}\right) \tag{2.49a}$$

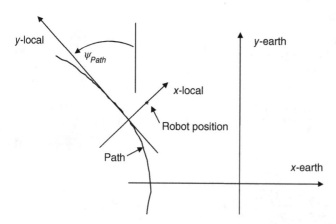

Figure 2.10 Path being followed by mobile robot and definition of local coordinate frame.

and

$$y_{path} = y_{path}\left(s_{path}\right)$$ (2.49b)

where

$$ds_{path}/dt = \sqrt{\dot{x}^2_{path} + \dot{y}^2_{path}}$$ (2.49c)

and

$$\tan \psi_{path} = -\dot{x}_{path}/\dot{y}_{path}$$ (2.49d)

The local coordinates of the robot are given by a rotation transformation as

$$\begin{bmatrix} x \\ y \end{bmatrix}_{local\ coords} = \begin{bmatrix} \cos \psi_{path} & \sin \psi_{path} \\ -\sin \psi_{path} & \cos \psi_{path} \end{bmatrix} \begin{bmatrix} x - x_{path} \\ y - y_{path} \end{bmatrix}_{earth\ coords}$$ (2.50)

or

$$x_{local} = \left[x - x_{path}\right] \cos \psi_{path} + \left[y - y_{path}\right] \sin \psi_{path}$$

$$y_{local} = -\left[x - x_{path}\right] \sin \psi_{path} + \left[y - y_{path}\right] \cos \psi_{path}$$

and

$$\psi_{local} = \psi - \psi_{path}$$

Now by the definition of the coordinate system, y_{local} is zero, i.e.,

$$-\left[x - x_{path}\right] \sin \psi_{path} + \left[y - y_{path}\right] \cos \psi_{path} = 0$$

Further, since it is always zero, its derivative with respect to time is also zero, i.e.,

$$d/dt\left\{ -\left[x - x_{path}\right] \sin \psi_{path} + \left[y - y_{path}\right] \cos \psi_{path} \right\} = 0$$

Evaluating the above differential and using the expressions for $\dot{x}$ and $\dot{y}$ as well as the fact that $\dot{x}_{path}$ may be written as $\dot{x}_{path} = -\dot{y}_{path} \tan \psi_{path}$ permits one to solve for $\dot{y}_{path}$.

$$\dot{y}_{path} = \left(V \cos \psi_{local} - x_{local}\dot{\psi}_{path}\right) \cos \psi_{path}$$ (2.51a)

From this result one can then obtain $\dot{x}_{path}$ as

$$\dot{x}_{path} = -\left(V \cos \psi_{local} - x_{local}\dot{\psi}_{path}\right) \sin \psi_{path}$$ (2.51b)

Now differentiating the equation for x_{local} yields

$$\dot{x}_{local} = \left[\dot{x} - \dot{x}_{path}\right] \cos \psi_{path} + \left[\dot{y} - \dot{y}_{path}\right] \sin \psi_{path}$$
$$- \left[x - x_{path}\right] \sin \psi_{path} \dot{\psi}_{path} + \left[y - y_{path}\right] \cos \psi_{path} \dot{\psi}_{path}$$

which is recognized as

$$\dot{x}_{local} = \left[\dot{x} - \dot{x}_{path}\right] \cos \psi_{path} + \left[\dot{y} - \dot{y}_{path}\right] \sin \psi_{path} - y_{local} \dot{\psi}_{path}$$

or

$$\dot{x}_{local} = \left[\dot{x} - \dot{x}_{path}\right] \cos \psi_{path} + \left[\dot{y} - \dot{y}_{path}\right] \sin \psi_{path}$$

since the definition of the coordinate system guarantees $y_{local} = 0$. Now again using the definitions for $\dot{x}$ and $\dot{y}$ the above can be expressed as

$$\dot{x}_{local} = -V \sin \psi \cos \psi_{path} - \dot{x}_{path} \cos \psi_{path} + V \cos \psi \sin \psi_{path} - \dot{y}_{path} \sin \psi_{path}$$

or

$$\dot{x}_{local} = -V \sin \psi_{local} - \dot{x}_{path} \cos \psi_{path} - \dot{y}_{path} \sin \psi_{path}$$

and since

$$\dot{x}_{path} = -\dot{y}_{path} \tan \psi_{path}$$

this reduces to simply

$$\dot{x}_{local} = -V \sin \psi_{local} \tag{2.52}$$

For the equation regarding the angle of the path we have

$$\dot{\psi}_{local} = \frac{V}{L} \tan \alpha - \dot{\psi}_{path} \tag{2.53}$$

Here we may express $\dot{\psi}_{path}$ in terms of its dependence on its location along the path and the motion of the coordinate origin along the path, i.e.,

$$\dot{\psi}_{path} = \frac{\partial \psi_{path}}{\partial s_{path}} \frac{ds_{path}}{dt}$$

It is recognized that $\partial \psi_{path}/\partial s_{path}$ is really the definition of curvature of the path, which may be designated as $\kappa(s_{path})$. The equation for ds_{path}/dt may be evaluated from its definition (2.49c) and is found through the use of (2.51) to be

$$\dot{s}_{path} = V \cos \psi_{local} - x_{local} \dot{\psi}_{path}$$

Thus

$$\dot{\psi}_{path} = \kappa(s_{path}) \left(V \cos \psi_{local} - x_{local} \dot{\psi}_{path} \right)$$

which may be rearranged and solved for $\dot{\psi}_{path}$ yielding

$$\dot{\psi}_{path} = \frac{\kappa(s_{path})}{1 + \kappa(s_{path})x_{local}} V \cos \psi_{local}$$

The equations for this model finally become

$$\dot{s}_{path} = \frac{1}{1 + \kappa(s_{path})x_{local}} V \cos \psi_{local} \qquad (2.54a)$$

$$\dot{x}_{local} = - V \sin \psi_{local} \qquad (2.54b)$$

and

$$\dot{\psi}_{local} = V \left(\frac{1}{L} \tan \alpha - \frac{\kappa(s_{path})}{1 + \kappa(s_{path})x_{local}} \cos \psi_{local} \right) \qquad (2.54c)$$

These three variables—displacement of the robot from the path, distance of the coordinate origin along the path, and heading of the robot with respect to the path—completely describe the kinematic state of the robot. From equation (2.54a) it is seen that $\dot{s}_{path}$ may either be greater than or less than the robot velocity V depending on whether the robot takes the inside or the outside when negotiating a curve. If the robot stays exactly on the path, then $\dot{s}_{path}$ and V are equal.

On the right-hand side of the last equation, the first term inside the parentheses represents the curvature of the robot trajectory. The allowable range of this term represents the robot steerability. The second term represents the effective curvature of the path and may be thought of as a disturbance. It will be assumed that the robot does have the steerability to negotiate the curves along the path when it starts on the path and aligned with it; otherwise, the control task would be impossible. Thus the allowable range of the steerability term must always exceed the allowable range of the disturbance term under the conditions that x_{local} is zero and that ψ_{local} is zero. The assumption regarding steerability then becomes

$$\frac{1}{L} \tan \alpha_{max} \geq \max \{ abs(\kappa(s_{path})) \} \qquad (2.55a)$$

or

$$\tan(\alpha_{max}) \geq L\kappa_{max} \qquad (2.55b)$$

It is instructive to also express these differential equations in terms of distance traveled by the robot, s. Since for any w

$$dw/ds = \frac{dw/dt}{ds/dt} = \frac{dw/dt}{V}$$

we obtain

$$ds_{path}/ds = s'_{path} = \frac{1}{1 + \kappa(s_{path})x_{local}} \cos \psi_{local} \qquad (2.56a)$$

$$dx_{local}/ds_{local} = x'_{local} = - \sin \psi_{local} \qquad (2.56b)$$

and similarly

$$d\psi_{local}/ds = \psi'_{local} = \frac{1}{L} \tan \alpha - \frac{\kappa(s_{path})}{1 + \kappa(s_{path})x_{local}} \cos \psi_{local} \qquad (2.56c)$$

These equations may be used in the study of steering control algorithms. One major accomplishment of this conversion to local coordinates is that the control problem has been transformed from a tracking problem into a regulator problem with a disturbance. For either formulation, a perfect solution requires future knowledge of the path curvature.

For the case of zero path curvature, the system of kinematic equations take on the familiar form

$$\dot{s}_{path} = V \cos \psi_{local} \qquad (2.57a)$$

$$\dot{x}_{local} = - V \sin \psi_{local} \qquad (2.57b)$$

and

$$\dot{\psi}_{local} = \frac{V}{L} \tan \alpha \qquad (2.57c)$$

and s_{path} becomes the local y axis. In terms of distance traveled, these equations are

$$s'_{path} = \cos \psi_{local} \qquad (2.58a)$$

$$x'_{local} = - \sin \psi_{local} \qquad (2.58b)$$

and

$$\psi'_{local} = \frac{1}{L} \tan \alpha \qquad (2.58c)$$

Example 2 *Explore the possibility of using the x_{local} and x'_{local} coordinates for control design.*

Solution 2

As a first step toward designing a controller for this system consider the case of zero path curvature above. By differentiating the equation for x'_{local} one has

$$x''_{local} = -\cos\psi_{local}\psi'_{local} = -\frac{\cos\psi_{local}}{L}\tan\alpha$$

A control is selected which depends on heading error and displacement from the path,

$$\tan\alpha = \frac{L}{\cos\psi_{local}}(-K_1\sin\psi_{local} + K_2 x_{local})$$

Here it appears that positive feedback has been used, but it is really negative. Recall that x is defined positive to the right while α and ψ are defined positive counter-clockwise. This expression written in terms of x_{local} becomes

$$\tan\alpha = \frac{L}{\cos\psi_{local}}(K_1 x'_{local} + K_2 x_{local})$$

leading to

$$x''_{local} + K_1 x'_{local} + K_2 x_{local} = 0$$

Clearly one could choose K_1 and K_2 to achieve any desired response in x_{local} as long as the equation for α yields a value within the achievable limits, i.e., abs(α) ≤ α_{max}. The effect of going to a second-order equation in x_{local} has been to map ψ_{local} into x'_{local}. Unfortunately, while things look fine in terms of x_{local} and x'_{local}, the equilibrium point $x_{local} = 0$ and $x'_{local} = 0$ maps into $x_{local} = 0$ and $\psi_{local} = n\pi$. For n odd this means the robot is in stable equilibrium but headed in the wrong direction. Thus, working in the x_{local}, x'_{local} space is not advisable for the case of large initial errors because of the one-to-many-mapping when going back to the x_{local}, ψ_{local} space.

A heuristic approach is now presented as a candidate for maintaining the mobile robot on path. This approach is to specify a profile for desired behavior for driving the error in displacement from the path (x_{local}) to zero, and then to allocate the remaining allowable velocity toward motion along the path ($\dot{s}_{path}$). It is arbitrarily specified that the velocity perpendicular to the path be

$$\dot{x}_{local-desired} = -\beta x_{local} \quad for \quad |\beta x_{local}| \leq V \tag{2.59a}$$

and

$$\dot{x}_{local-desired} = -V\,\text{sign}(x_{local}) \quad for \quad |\beta x_{local}| > V \tag{2.59b}$$

Using the remaining available velocity, $\sqrt{V^2 - \dot{x}^2_{local-desired}}$, for motion along the path implies

$$\tan(\psi_{local-desired}) = \frac{-\dot{x}_{local-desired}}{\sqrt{V^2 - \dot{x}^2_{local-desired}}} \tag{2.60a}$$

or

$$\psi_{local-desired} = \tan^{-1} \frac{-\dot{x}_{local-desired}}{\sqrt{V^2 - \dot{x}^2_{local-desired}}} \qquad (2.60b)$$

or

$$\psi_{local-desired} = -\sin^{-1} \frac{\dot{x}_{local-desired}}{V} \qquad (2.60c)$$

It may be seen from the first form that this equation for desired heading guarantees

$$-\pi/2 \le \psi_{local-desired} \le \pi/2$$

and thus ensures that the desired component of velocity along the path will be in the proper direction. One then steers the robot to the desired heading $\psi_{local-desired}$ as follows

$$\dot{\psi}_{local} = \gamma(\psi_{local-desired} - \psi_{local}) \qquad (2.61a)$$

or

$$\frac{V}{L} \tan \alpha = \gamma(\psi_{local-desired} - \psi_{local}) + V\left(\frac{\kappa}{1 + \kappa x_{local}}\right) \cos \psi_{local} \qquad (2.61b)$$

or

$$\alpha = \tan^{-1}\left\{\frac{L}{V}\gamma(\psi_{local-desired} - \psi_{local}) + L\left(\frac{\kappa}{1 + \kappa x_{local}}\right) \cos \psi_{local}\right\} \quad \text{for}$$

$$\left|\frac{L}{V}\gamma(\psi_{local-desired} - \psi_{local}) + L\left(\frac{\kappa}{1 + \kappa x_{local}}\right) \cos \psi_{local}\right| \le \tan \alpha_{max}$$

$$(2.61c)$$

and

$$\alpha = \alpha_{max}\text{sign}\left\{\frac{L}{V}\gamma(\psi_{local-desired} - \psi_{local}) + L\left(\frac{\kappa}{1 + \kappa x_{local}}\right) \cos \psi_{local}\right\}\text{for}$$

$$\left|\frac{L}{V}\gamma(\psi_{local-desired} - \psi_{local}) + L\left(\frac{\kappa}{1 + \kappa x_{local}}\right) \cos \psi_{local}\right| > \tan \alpha_{max}$$

$$(2.61d)$$

In implementing such a control one must always check to see whether

$$|\psi_{local-desired} - \psi_{local}| < \pi$$

and if not, then add or subtract 2π as many times as is needed for this to be true.

When the controller is operating in the linear region, i.e., not saturated, the equation (2.61a) may be manipulated to become

$$\dot{\psi}_{local} = \gamma\left(-\sin^{-1}\left(\frac{\dot{x}_{local-desired}}{V}\right) - \psi_{local}\right)$$

or

$$\dot{\psi}_{local} = \gamma\left(\sin^{-1}\left(\frac{\beta x_{local}}{V}\right) - \sin^{-1}\left(\frac{-\dot{x}_{local}}{V}\right)\right)$$

and

$$\dot{x}_{local} = -V\sin\left(\psi_{local}\right)$$

Linearizing the sine and inverse sine functions yields

$$\dot{\psi}_{local} = \gamma\left(\frac{\beta x_{local}}{V} + \frac{\dot{x}_{local}}{V}\right) \tag{2.62a}$$

and

$$\dot{x}_{local} = -V\psi_{local} \tag{2.62b}$$

which may be combined to yield

$$\ddot{x}_{local} + \gamma\dot{x}_{local} + \gamma\beta x_{local} = 0 \tag{2.63}$$

The above is guaranteed to be stable for all positive values of β and γ. A crucial difference in this control strategy and the one presented in the preceding example is that in the preceding example the tangent of α depended on the sine of the heading error, while here the tangent of α depends on the heading error itself.

Equation (2.61) is another instance of applying computed control. The first term in equation (2.61b) provides linear feedback and the second term cancels out the nonlinear disturbance term (a curving path). Here one could select the parameters β and γ corresponding to the desired closed-loop system behavior. To implement such a controller requires that the commanded steering angle, α not exceed its allowable range. It is expected that the left side of equation (2.61b) could be made equal to the second term on the right side, i.e., steerability must exceed the disturbance. However, care must be exercised in selecting β and γ so that the total of all the terms on the right side can still be satisfied by the left side. Otherwise the controller operates in the saturated mode and the closed-loop dynamics are not the same as those predicted by the linear equations.

Implementation also requires that one be able to sense y_{local}, ψ_{local}, and κ. Sensing the displacement from the path, y_{local} and the current heading of the robot relative to the path, ψ_{local} would be required even for a straight path. What is required additionally here is κ, the curvature of the path at the current robot location. Sensing of portions of the path not yet traversed and estimating the curvature from these observations would be required.

If one chooses to use the computed control without knowledge of the path curvature, i.e., according to

$$\alpha = \tan^{-1}\left\{\frac{L}{V}\gamma\left(\psi_{local-desired} - \psi_{local}\right)\right\}$$

where

$$\psi_{local-desired} = \sin^{-1}\frac{\dot{x}_{local-desired}}{V} = \sin^{-1}\frac{-\beta x_{local}}{V}$$

when, in fact, the path does have curvature, the linearized closed-loop equations become

$$\ddot{x}_{local} + \gamma\dot{x}_{local} + \gamma\beta x_{local} = -V^2\frac{\kappa\left(s_{path}\right)}{1 + \kappa\left(s_{path}\right)x_{local}}\cos\psi_{local}$$

The solution is now influenced by the disturbance on the right side of the equation which can cause transient as well as steady-state errors.

A simulation of the system using this control strategy is presented in Figure 2.11. Here the initial condition of the robot is $x = 10$, $y = 0$, and $\psi = -\pi/2$ with $\beta = 2.0$ and $\gamma = 0.5$. The path is a circle of radius 8. Observe that the robot initially turns as steeply as possible and then briefly heads to the

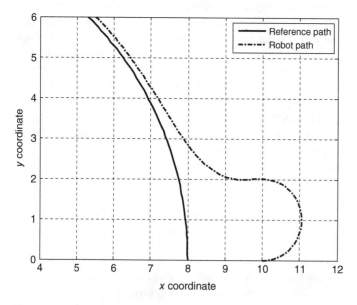

Figure 2.11 Simulation A illustrating the heuristic steering control strategy.

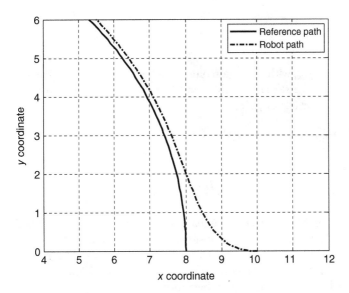

Figure 2.12 Simulation B illustrating the heuristic steering control strategy.

closest point on the path. It then merges with the path according to equation (2.60). Because of the path curvature and the fact that the control assumed zero curvature, there is a tracking error. This error would remain as long as the path has this curvature.

A second simulation is shown in Figure 2.12. Here $x = 11$, $y = 0$, and $\psi = \pi/2$. Note the first portion shows motion directly toward the path followed by a gradual turn to move along the path.

Classical linear control techniques may be used here if one uses linearized equations and assumes that V will be constant. Here we shall use the equations in the time domain. By making the following approximations

$$\tan \alpha \approx \alpha$$
$$\sin \psi_{local} \approx \psi_{local}$$
$$\cos \psi_{local} \approx 1$$

and

$$\frac{\kappa(s)}{1 + \kappa(s)x_{local}} \cos \psi_{local} \approx \kappa$$

the differential equations of interest become

$$\dot{\psi}_{local} = \frac{V}{L}\alpha - V\kappa$$

and

$$\dot{x}_{local} = -V\psi_{local}$$

Now feeding back $-x_{local}$ with unity gain plus K_1 times the local heading angle, ψ_{local} and then setting the steering angle α equal to a combination of proportional plus integral error terms yields

$$\alpha(s) = \left(K_p + K_i/s\right)\left(ref + x_{local}(s) - K_1\psi_{local}(s)\right)$$

or

$$\alpha(s) = \left(K_p + K_i/s\right)\left[ref + (1 + K_1s/V)x_{local}(s)\right].$$

The above may appear to be positive feedback, but it is not. As was discussed in a previous example, recall that x_{local} is positive to the right, and that ψ is positive in the counter-clockwise direction. The corresponding block diagram in local coordinates with the feedback control is shown in Figure 2.13.

The closed-loop transfer function then becomes

$$x_{local}(s) = \frac{\left(K_pV^2/L\right)s + \left(V^2K_i/L\right)}{s^3 + \left(VK_pK_1/L\right)s^2 + \left(V^2K_p/L + VK_1K_i/L\right)s + V^2K_i/L}ref$$
$$+ \frac{Vs}{s^3 + \left(VK_pK_1/L\right)s^2 + \left(V^2K_p/L + VK_1K_i/L\right)s + V^2K_i/L}\kappa$$

The steady-state gain of the first portion is seen to be unity guaranteeing good steady-state reference tracking. The s in the numerator of the second portion guarantees zero steady-state error from the disturbance. Finally, the choice of the three gain parameters, K_pK_i and K_1 will dictate the locations of the closed-loop poles.

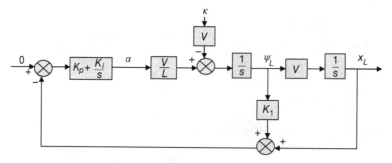

Figure 2.13 Block diagram of the closed-loop steering system in local coordinates.

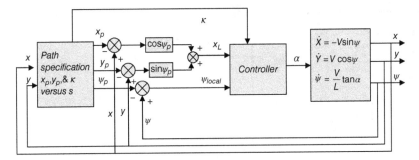

Figure 2.14 Block diagram in original coordinates for mobile robot following a path.

In Figure 2.14 is a block diagram in terms of original coordinates showing how the local coordinates are obtained. Here x, y, and κ are assumed to be functions of the distance traveled along the path, s. The operations represented here have application for any steering control algorithm which is computed based on the local coordinates.

The nonlinear robot equations for the robot have been simulated utilizing the control algorithm based on classical linear control theory. The behavior was quite good as long as the initial deviation from the path did not exceed several meters and the curvature of the path did not exceed that of the robot's capability, $\tan(\alpha_{max})/L$. Some illustrative examples follow and are shown in Figures 2.15–2.17. Varying responses may be obtained by adjusting the gain parameters. Here the gains were set at $K_1 = 4.618$, $K_p = 0.9527$, and $K_i = 0.4$. The resulting closed-loop poles for the linearized system were at -1, -1, and -0.2. Note that there is a finite zero in the closed-loop transfer function that could cause overshoot in some cases even though the roots of the denominator correspond to an overdamped system.

2.11 Optimal Steering of Front-Wheel Steered Vehicle

The objective of the previous control algorithms was to provide control action that would stabilize the overall behavior and yield good performance. In this section, a different approach is taken. It will be assumed that the robot will operate at a fixed velocity and the objective is to steer it in such a way that it will reach the destination in minimum time. It is further assumed that there are no obstacles so that the robot is free to travel anywhere without being confined to a roadway. Note that the minimum-time trajectory would also be the minimum-distance trajectory since speed is fixed. The proceeding builds on the theory of optimal control. The interested reader may consult appropriate references for more background on this body of knowledge. The equations of motion are repeated once more for convenience.

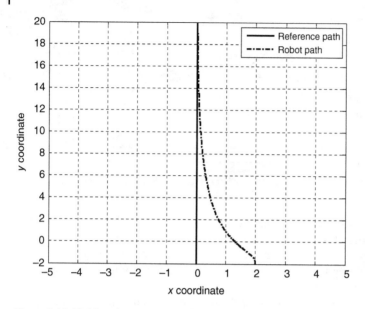

Figure 2.15 Mobile robot recovering from 2 m error while tracking a straight line.

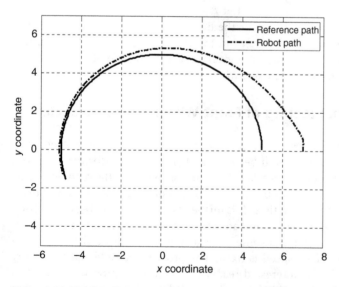

Figure 2.16 Mobile robot recovering from 2 m error while tracking a segment of a circle of radius 5 m.

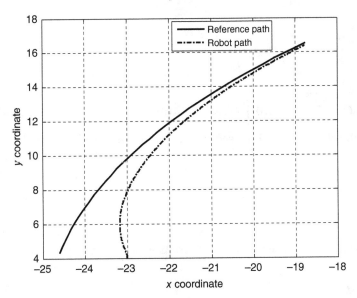

Figure 2.17 Mobile robot recovering from 2 m error while tracking a gently sloping curve.

$$\dot{x} = -V \sin \psi$$
$$\dot{y} = V \cos \psi$$

and

$$\dot{\psi} = \frac{V}{L} \tan \alpha$$

We now form the Hamiltonian

$$H = 1 - \lambda_x V \sin \psi + \lambda_y V \cos \psi + \lambda_\psi \frac{V}{L} \tan \alpha \qquad (2.64)$$

The equations for the co-states become

$$\dot{\lambda}_x = -\frac{\partial H}{\partial x} = 0$$

or

$$\lambda_x = C_1 \qquad (2.65a)$$

$$\dot{\lambda}_y = -\frac{\partial H}{\partial y} = 0$$

or

$$\lambda_y = C_2 \qquad (2.65b)$$

and

$$\dot{\lambda}_\psi = -\frac{\partial H}{\partial \psi} = -\lambda_x V \cos \psi + \lambda_y V \sin \psi = -\lambda_x \dot{y} - \lambda_y \dot{x}$$

or

$$\lambda_\psi(t) = C_3 - C_1\{y(t) - y(0)\} - C_2\{x(t) - x(0)\} \tag{2.65c}$$

Examining the Hamiltonian, it is seen that it will be minimized by choosing

$$\alpha = -\alpha_{max}\mathrm{sign}(\lambda_\psi)$$

whenever

$$\lambda_\psi \neq 0$$

As was shown earlier, this control action results in segments of circles with radius

$$R = \frac{L}{\tan \alpha_{max}}$$

Thus, portions of the optimal trajectory are segments of circles. It is of interest to examine the possibility of singular control, i.e., what if $\lambda_\psi = 0$? To answer this we note that for $\lambda_\psi = 0$ over a nonzero interval we must also have

$$\dot{\lambda}_\psi = 0$$

or

$$-\lambda_x \dot{y} - \lambda_y \dot{x} = 0$$

or

$$-C_1\dot{y} - C_2\dot{x} = 0 \tag{2.66a}$$

which implies that the robot is moving in a straight line with slope

$$m = -\frac{C_2}{C_1} \tag{2.66b}$$

This straight-line motion can happen if and only if $\alpha = 0$. Thus the optimal control obeys the following

$$\alpha = -\alpha_{max}\mathrm{sign}(\lambda_\psi); \ \lambda_\psi \neq 0 \tag{2.67a}$$

and

$$\alpha = 0; \ \lambda_\psi = 0 \tag{2.67b}$$

At this point it can be concluded that the optimal trajectory is seen to be a series of segments that are either segments of clockwise circles, segments of counter-clockwise circles, or straight lines. The equation developed for λ_ψ, i.e.,

$$\lambda_\psi(t) = C_3 - C_1\{y(t) - y(0)\} - C_2\{x(t) - x(0)\} \tag{2.68a}$$

can also be expressed in terms of the final conditions as

$$\lambda_\psi(t) = D_3 + C_1\{y(t_f) - y(t)\} + C_2\{x(t_f) - x(t)\} \tag{2.68b}$$

By solving for $\lambda_\psi(t_f)$ using each of the expressions it is easy to show that

$$D_3 = C_3 - C_1\{y(t_f) - y(0)\} - C_2\{x(t_f) - x(0)\} \tag{2.69}$$

The first form is useful in working forward from the initial conditions to determine where the trajectory switches from a circular arc

$$\alpha = \pm\alpha_{max}$$

to a singular arc

$$\alpha = 0$$

The second form is useful for working backward from the final conditions to determine where the trajectory switches from a singular arc

$$\alpha = 0$$

to a circular arc

$$\alpha = \pm\alpha_{max}$$

Note that the problem is a two-point boundary value problem as would be expected for an optimal control problem. The three specified final conditions on x, y, and ψ provide the constraining equations that determine the unknown initial conditions, C_1, C_2, and C_3 for the co-state variables. These in turn determine D_3 as per the above equation. The test for determining whether the values for C_1, C_2, and C_3 are the correct ones for the particular problem at hand is to integrate the equations of motion and the co-state equations forward in time from the initial conditions until the beginning of the singular arc, i.e., λ_ψ becomes zero. Now using these same values for C_1, C_2, and C_3, and the boundary conditions on x, y, and ψ, one determines D_3 and then integrates the equations of motion backward from the specified final conditions until the end of the singular arc, i.e., again λ_Φ becomes zero. The conditions at the beginning (arrive) and end (depart) of the singular arc should satisfy the equation

$$-C_1\left(y_{arrive} - y_{depart}\right) - C_2\left(x_{arrive} - x_{depart}\right) = 0 \tag{2.70}$$

This is equivalent to the earlier equation

$$-C_1\dot{y} - C_2\dot{x} = 0$$

along the singular arc.

Several examples will now be presented to illustrate how the unknown coefficients may be determined. First the radius of curvature when using maximum steering angle will be evaluated. Assume that the length of the mobile robot is 2 and that the maximum steering angle is $\pi/4$. Then evaluating the equation for the radius of curvature we obtain

$$R = \frac{L}{\tan \alpha_{max}} = 2$$

Example 3   *Take as the initial conditions*

$$[x(0) \quad y(0) \quad \psi(0)]^T = [-20 \quad -4 \quad 0]^T$$

and as the final conditions

$$[x(t_f) \quad y(t_f) \quad \psi(t_f)]^T = [0 \quad 0 \quad 0]^T$$

As will be seen, the minimum-time trajectory is comprised of a 90° section of a clockwise circle followed by a straight line of length 16 followed by a 90° section of a counter-clockwise circle.

Solution 3

Utilizing the boundary conditions for the problem coupled with the conditions

$$H(t_f) = 0$$

$$\frac{dH}{dt} = 0$$

and along the singular arc

$$\lambda_\psi = 0$$

yields as solutions for the co-state parameters

$$[C_1 \quad C_2 \quad C_3 \quad D_3]^T = \left[\frac{L}{2V} \quad 0 \quad \frac{L}{V} \quad -\frac{L}{V}\right]^T$$

Thus

$$\lambda_\psi = \frac{L}{V} - \frac{L}{2V}\{y(t) - y(0)\}$$

is used at the beginning of the trajectory to determine where the singular arc begins, and

$$\lambda_\psi = -\frac{L}{V} + \frac{L}{2V}\{y(t_f) - y(t)\}$$

is used at the end of the trajectory to determine where the singular arc ends.

The fact that $-C_2/C_1 = 0$ *is consistent with the fact that the slope of the singular arc of the trajectory is zero.*

Plots of y versus x, λ_ψ *versus t, and* α *versus t follow in Figure 2.18a–c.*

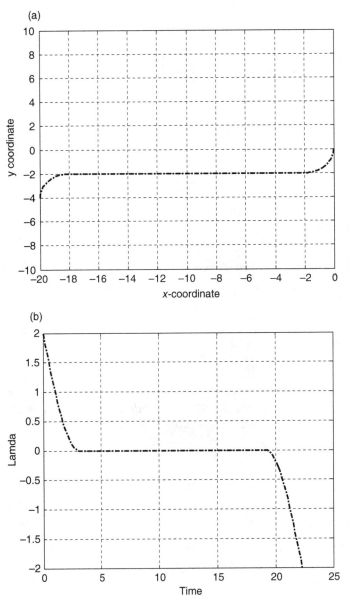

Figure 2.18 (a) Trajectory in x–y space. (b) λ_ψ versus time. (c) α versus time for Example 3.

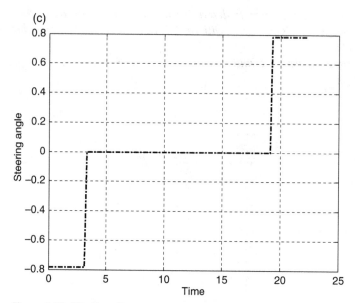

Figure 2.18 (Continued)

Example 4 *Take as the initial conditions*

$$[x(0) \quad y(0) \quad \psi(0)]^T = [-20 \quad 0 \quad -\pi]^T$$

and as the final conditions

$$[x(t_f) \quad y(t_f) \quad \psi(t_f)]^T = [0 \quad 0 \quad 0]^T$$

Solution 4

The minimum-time trajectory is comprised of a 90° section of a counter-clockwise circle followed by a straight line of length 16 followed by a 90° section of a counter-clockwise circle. The solutions for the co-states yield

$$[C_1 \quad C_2 \quad C_3 \quad D_3]^T = \left[\frac{L}{2V} \quad 0 \quad -\frac{L}{V} \quad -\frac{L}{V} \right]^T$$

Thus,

$$\lambda_\psi = -\frac{L}{V} - \frac{L}{2V}\{y(t) - y(0)\}$$

is used at the beginning of the trajectory, and

$$\lambda_\psi = -\frac{L}{V} + \frac{L}{2V}\{y(t_f) - y(t)\}$$

is used at the end of the trajectory.

The fact that $-C_2/C_1 = 0$ *is consistent with the fact that the slope of the singular arc of the trajectory is zero. Plots of y versus x,* λ_ψ *versus t, and* α *versus t follow in Figure 2.19a–c.*

(a)

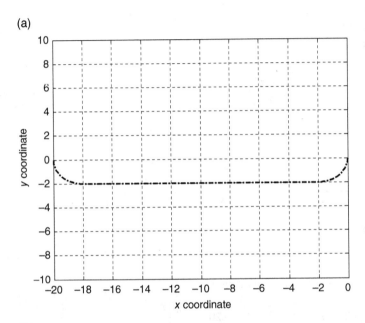

(b)

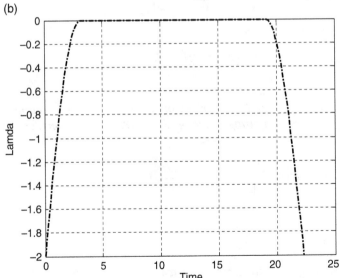

Figure 2.19 (a) Trajectory in x–y space. (b) λ_ψ versus time. (c) α versus time for Example 4.

(c)

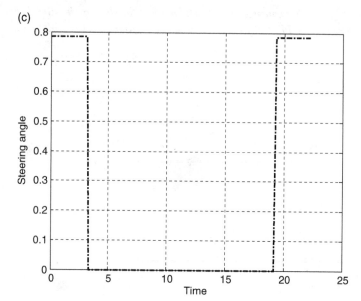

Figure 2.19 (Continued)

Example 5 *Take as the initial conditions*

$$\begin{bmatrix} x(0) & y(0) & \psi(0) \end{bmatrix}^T = \begin{bmatrix} -2 & -20 & -\dfrac{\pi}{2} \end{bmatrix}^T$$

and as the final conditions

$$\begin{bmatrix} x(t_f) & y(t_f) & \psi(t_f) \end{bmatrix}^T = \begin{bmatrix} 0 & 0 & 0 \end{bmatrix}^T$$

Solution 5

The minimum-time trajectory is comprised of a 90° section of a counter-clockwise circle followed by a straight line of length 18. The solutions for the co-states yield

$$\begin{bmatrix} C_1 & C_2 & C_3 & D_3 \end{bmatrix}^T = \begin{bmatrix} 0 & -\dfrac{L}{V} & -\dfrac{2L}{V} & 0 \end{bmatrix}^T$$

Thus

$$\lambda_\psi = -\dfrac{2L}{V} + \dfrac{L}{V}\{x(t) - x(0)\}$$

is used at the beginning of the trajectory. Here it is not necessary to describe the equation for λ_ψ in terms of the terminal conditions since the final portion of the *trajectory is a singular arc. The fact that* $-C_1/C_2 = 0$ *is consistent with the fact that the slope of the singular arc of the trajectory is infinite. Plots of y versus x, λ_ψ versus t, and α versus t follow in Figure 2.20a–c.*

(a)

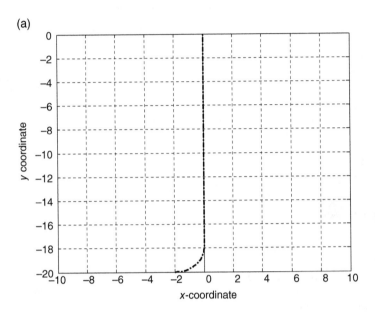

(b)

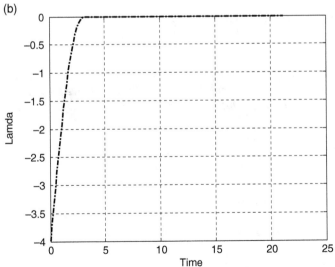

Figure 2.20 (a) Trajectory in x–y space. (b) λ_ψ versus time. (c) α versus time for Example 5.

(c)

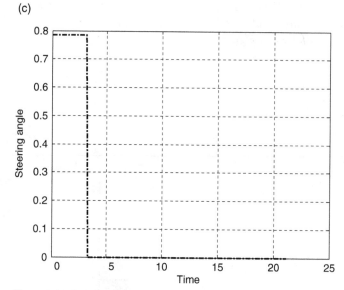

Figure 2.20 (Continued)

Example 6   *Take as the initial conditions*

$$[x(0) \quad y(0) \quad \psi(0)]^T = \begin{bmatrix} -4 & -20 & -\dfrac{\pi}{2} \end{bmatrix}^T$$

and as the final conditions

$$[x(t_f) \quad y(t_f) \quad \psi(t_f)]^T = \begin{bmatrix} 0 & 0 & -\dfrac{\pi}{2} \end{bmatrix}^T$$

Solution 6

The minimum-time trajectory is comprised of a 90° section of a counter-clockwise circle followed by a straight line of length 16 followed by a 90° section of a clockwise circle. The solutions for the co-states yield

$$[C_1 \quad C_2 \quad C_3 \quad D_3]^T = \begin{bmatrix} 0 & -\dfrac{L}{2V} & -\dfrac{L}{V} & \dfrac{L}{V} \end{bmatrix}^T$$

Thus

$$\lambda_\psi = -\frac{L}{V} + \frac{L}{2V}\{x(t) - x(0)\}$$

is used at the beginning of the trajectory, and

$$\lambda_\psi = \frac{L}{V} - \frac{L}{2V}\{x(t_f) - x(t)\}$$

is used at the end of the trajectory.

The fact that $-C_1/C_2 = 0$ *is consistent with the fact that the slope of the singular arc of the trajectory is infinite. Plots of y versus x, λ_ψ versus t, and α versus t follow in Figure 2.21a–c.*

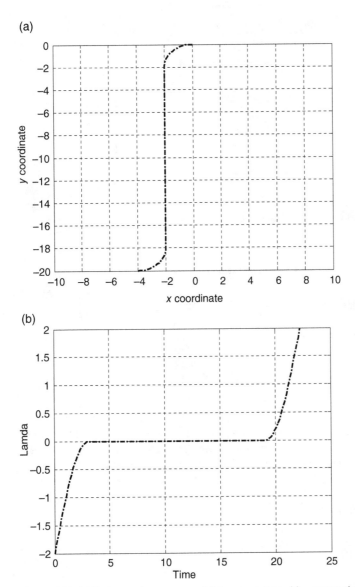

Figure 2.21 (a) Trajectory in x–y space. (b) λ_ψ versus time. (c) α versus time for Example 6.

(c)

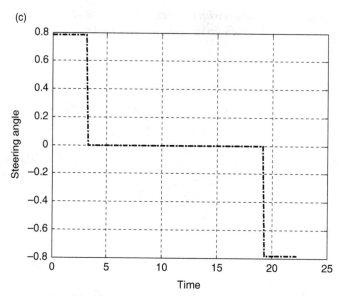

Figure 2.21 (Continued)

Example 7 *Take as the initial conditions*

$$[x(0) \quad y(0) \quad \psi(0)]^T = \left[-24\sqrt{0.5}, \quad -4-16\sqrt{0.5}, \quad -\frac{\pi}{2} \right]^T$$

and as the final conditions

$$[x(t_f) \quad y(t_f) \quad \psi(t_f)]^T = \left[0 \quad 0 \quad -\frac{\pi}{2} \right]^T$$

Solution 7

The minimum-time trajectory is comprised of a 45° section of a counter-clockwise circle followed by a straight line of length 20 followed by a 45° section of a clockwise circle. The solutions for the co-states yield

$$C_1 = \frac{L}{V} \frac{1}{2(\sqrt{2}-1) + L}$$

$$C_2 = \frac{-L}{V} \frac{1}{2(\sqrt{2}-1) + L}$$

$$C_3 = -\frac{L}{V} + \frac{L^2}{V} \frac{1}{2(\sqrt{2}-1) + L}$$

and

$$D_3 = \frac{L}{V} - \frac{L^2}{V} \frac{1}{2(\sqrt{2}-1) + L}$$

For

$$L = 2$$

which is what was assumed here, these reduce to

$$[\,C_1 \quad C_2 \quad C_3 \quad D_3\,]^T = \left[\,\frac{1}{\sqrt{2}V} \quad -\frac{1}{\sqrt{2}V} \quad \frac{-2+\sqrt{2}}{V} \quad \frac{2-\sqrt{2}}{V}\,\right]^T$$

Thus

$$\lambda_\psi = \frac{-2+\sqrt{2}}{V} - \frac{1}{\sqrt{2}V}(y(t) - y(0)) + \frac{1}{\sqrt{2}V}\{x(t) - x(0)\}$$

is used at the beginning of the trajectory, and

$$\lambda_\psi = \frac{2-\sqrt{2}}{V} + \frac{1}{\sqrt{2}V}(y(t_f) - y(t)) - \frac{1}{\sqrt{2}V}\{x(t_f) - x(t)\}$$

is used at the end of the trajectory.

The fact that $-C_2/C = 1$ is consistent with the fact that the slope of the singular arc of the trajectory is unity. Plots of y versus x, λ_ψ versus t, and α versus t follow in Figure 2.22a–c.

It would be desirable to obtain a control law which would provide the optimal steering control as a function of the present state. For a fixed set of final conditions, it is conceivable that such a control law does exist. However, the fact that the state space is of dimension three makes this a difficult problem, i.e., the surface for switching from maximum steering angle to zero steering angle and vice versa would be a surface described as a function of x, y, and ψ. Nevertheless, one can take advantage of the nature of the candidate segments of optimal trajectories as dictated by the necessary conditions and use geometrical reasoning to solve for the complete optimal trajectory given the boundary conditions. Optimal trajectories were seen to be a series of segments that are either arcs of clockwise circles, arcs of counter-clockwise circles or straight lines.

An example will be presented to illustrate this approach.

(a)

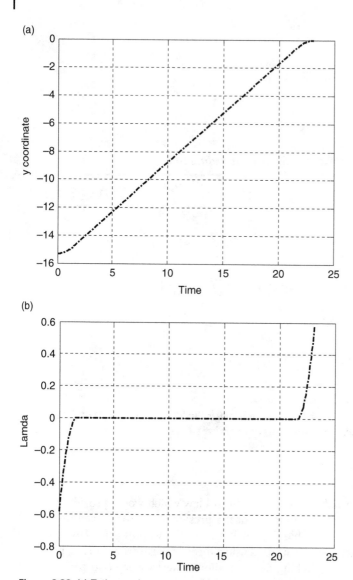

(b)

Figure 2.22 (a) Trajectory in *x–y* space. (b) λ_y versus time. (c) α versus time for Example 7.

(c)

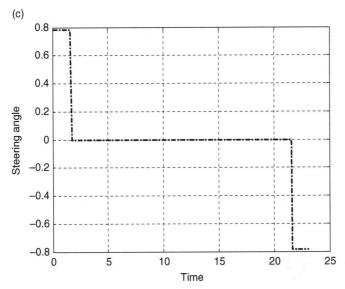

Figure 2.22 (Continued)

Example 8 *Take as the initial conditions*

$$[x(0) \quad y(0) \quad \psi(0)]^T = \left[-24\sqrt{0.5}, \quad -4 - 16\sqrt{0.5}, \quad -\frac{\pi}{2} \right]^T$$

and as the final conditions

$$[x(t_f) \quad y(t_f) \quad \psi(t_f)]^T = \left[0 \quad 0 \quad -\frac{\pi}{2} \right]^T$$

Next construct the possible trajectories emanating from the initial state as well as those terminating at the final state. These are sections of circles in the clockwise direction and sections of circles in the counter-clockwise direction, all at the minimum radius of curvature of the mobile robot, or sections of straight lines.

Now connect one circle from the initial state with one from the final state circles with a straight line that is tangent to each. By inspection one can decide which circles to connect so that the path is of minimum length. Figure 2.23 illustrates the optimal solution for this example. It is seen that the optimal trajectory consists of a brief turn in the counter clockwise direction where the steering angle is maximum to the left followed by a straight section where the steering angle is zero followed by a brief turn in the clockwise direction where the steering angle is maximum to the right.

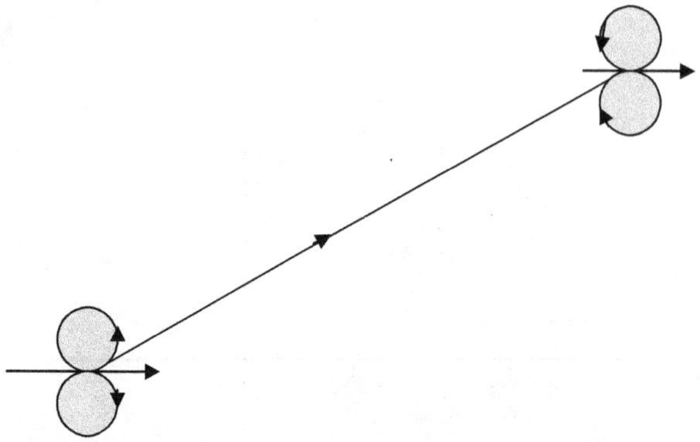

Figure 2.23 Trajectory in *x*–*y* space for Example 8.

Example 9 *Take as the initial conditions*

$$[x(0) \quad y(0) \quad \psi(0)]^T = [-5, \quad 0, \quad 3\pi/4]^T$$

and as the final conditions

$$[x(t_f) \quad y(t_f) \quad \psi(t_f)]^T = [0 \quad 0 \quad 0]^T$$

The same procedure as above is followed in this example. Figure 2.24 illustrates the optimal solution for this example. It is seen that the optimal trajectory consists of a segment of a turn in the counter clockwise direction where the steering angle is maximum to the left followed by a straight section where the steering angle is zero followed by another turn in the counter clockwise direction where the steering angle is again maximum to the left.

Figure 2.24 Trajectory in *x*–*y* space for Example 9.

2.12 Optimal Steering of Front-Wheel Steered Vehicle, Free Final Heading Angle

A simpler problem results for the case where the final heading angle is left free. For this case we have that the final value of the associated co-state variable is zero, i.e.,

$$\lambda_\psi(t_f) = 0$$

The net result of this is that the last segment of the optimal trajectory is singular, i.e., the steering angle for this segment is zero. The optimal solution is then an arc of the minimum-radius circle followed by a straight-line segment. The robot is steered at maximum steering angle until it is pointed at the final destination. Then it travels in a straight line.

For

$$\psi(t) > \tan^{-1}\left(\frac{-\{x(t_f) - x(t)\}}{\{y(t_f) - y(t)\}}\right) \tag{2.71a}$$

one uses

$$\alpha = -\alpha_{max} \tag{2.71b}$$

for

$$\psi(t) < \tan^{-1}\left(\frac{-\{x(t_f) - x(t)\}}{\{y(t_f) - y(t)\}}\right) \tag{2.72a}$$

one uses

$$\alpha = \alpha_{max} \tag{2.72b}$$

and for

$$\psi(t) = \tan^{-1}\left(\frac{-\{x(t_f) - x(t)\}}{\{y(t_f) - y(t)\}}\right) \tag{2.73a}$$

one uses

$$\alpha = 0 \tag{2.73b}$$

An illustration of an optimal trajectory with free final heading is shown in Figure 2.25. The robot travels in a circle of minimum radius using $\alpha = \pm\alpha_{max}$ until it is headed toward the destination. Then it travels in a straight line with $\alpha = 0$.

In summary, the necessary conditions for the optimal control of the mobile robot have been derived. For the case where the final heading and final position

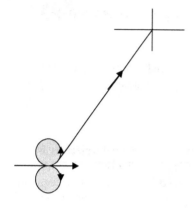

Figure 2.25 Trajectory in x–y space.

are specified, the final segment of the trajectory and the initial segment of the trajectory are sections of circles. These segments are connected by a straight line corresponding to singular control. Through a proper choice of the unknown parameters of the co-state variables, a solution for a particular example may be obtained. For the switches that must take between singular segments and nonsingular segments, one must always go from the nonsingular segments to the singular segments. An alternative and simpler approach uses geometrical reasoning coupled with the nature of the solution. Both methods are illustrated via examples. A closed-form solution has been obtained for the case where the final orientation is free.

Exercises

1   Use the linearized model for the front-wheel steered robot and take the y axis as the reference path, i.e., $y_{ref} = Vt$, $x_{ref} = 0$, and $\psi_{ref} = 0$. Use as control algorithms, $\delta V = -K_1\delta y$ and $\delta\alpha = K_2\delta x - K_3\delta\psi$. Determine the solutions for the closed-loop eigenvalues and find the conditions on the K' s for stable behavior.

2   Develop an algorithm for speed control and one for heading control. Do this for both type robots. Assume that you have measurements of position, heading, and velocity. Simulate for a step change in heading. Also simulate for a step change in speed. Modify your algorithms as needed for desirable behavior. Since the system is nonlinear, it is more difficult to determine whether it is stable than if it were linear. Simulate any conditions you wish to test for stability.

3 Repeat the above control design for the case where there is no direct measurement of either heading or speed. Use finite differences of the x and y measurements to approximate heading and speed. Simulate as before.

4 Assume the robot is located at $x = 3$ and $y = 5$. The desired location is $x = 8$ and $y = 35$. What is the instantaneous desired heading angle for the robot to move toward the target?

5 Consider the definition of a local coordinate system as described in the previous chapter. What are the measurements that one must have available in order to implement a control strategy based on such a coordinate system? What kind of approximations could one use to simplify these requirements? Can you think of any coordinate system and accompanying control strategy that would not require knowledge of displacement of the robot from the center of the lane?

6 Using computed control one can theoretically cause a system to behave in any desired manner. Discuss the practical limitations when using this method. What information is required in forming the control signal? What if one asks the system to perform beyond its physical capabilities, e.g., excessively fast or with an excessively small radius of curvature?

7 Solve the optimal robot steering problem when the final heading is free. The initial heading angle is $-\pi/2$ radians and the initial position is $x = 0$ and $y = 0$. The final position is $x = 2$ m and $y = 20$ m. The minimal radius of curvature for the robot is 2 m. The performance measure is distance traveled.

8 Use the results for the minimum-time solutions for the front-wheel steered robot and find the solution for the optimal trajectory when $y(0) = -12$, $y(t_f) = 0$, $x(0) = 0$, $x(t_f) = 0$, $\psi(t_0) = \pi/2$, and $\psi(t_f) = 0$.

References

Athans, M. and Falb, P. L., *Optimal Control: An Introduction to the Theory and Its Application*, McGraw-Hill, New York, 1966.

Brogan, W. L., *Modern Control Theory*, Prentice Hall, Upper Saddle River, NJ, 1991.

Bryson, A. E. and Ho, Y.-C., *Applied Optimal Control: Optimization, Estimation and Control*, Blaisdell, Waltham, MA, 1969.

Canudas de Wit, Carlos, Siciliano, Bruno, and Bastin, Georges (eds.), *Theory of Robot Control*, Springer, 1996.

De Luca, A., Oriolo, G., and Vendittelli, M., Control of Wheeled Mobile Robots: An Experimental Overview, Nicosia, S., Siciliano, B., Bicchi, A., and Valigi, P. (eds.), *Ramsete. Lecture Notes in Control and Information Sciences*, Vol. **270**, Springer, Berlin, Heidelberg, 2001, pp. 181–226.

Dixon, W. E., Dawson, D. M., Zergeroglu, E., and Behal, A., *Nonlinear Control of Wheeled Robots*, Springer-Verlag London Limited, 2001.

Dixon, W. E., Dawson, D. M., Zergeroglu, E., and Zhang, F., Robust Tracking and Regulation Control for Mobile Robots, *International Journal of Robust and Nonlinear Control*, Vol. **10** (2000), pp. 199–216.

Fahimi, Farbod, *Autonomous Robots: Modeling, Path Planning, and Control*, Springer, 2010.

Kansal, S., Jakkidi, S., and Cook, G., "The Use of Mobile Robots for Remote Sensing and Object Localization," *Proceedings of IECON 2003* (pp. 279–284, Roanoke, VA, November 2–6, 2003).

Lavalle, Steven M., *Planning Algorithms/Motion Planning*, Cambridge University Press, 2006.

Moon Kim, B. and Tsiotras, P., "Controllers for Unicycle-Type Wheeled Robots: Theoretical Results and Experimental Validation," *IEEE Transactions on Robotics and Automation*, Vol. **18**, No. 3 (June 2002), pp. 294–307.

Rusu, Radu Bogdan and Borodi, Marius, On Computing Robust Controllers for Mobile Robot Trajectory Calculus: Lyapunov, Unpublished paper series at http://citeseerx.ist.psu.edu/viewdoc/download?doi=10.1.1.526.8554&rep=rep1&type=pdf (accessed September 13, 2019).

3

Robot Attitude

3.1 Introduction

This chapter is devoted to the introduction of coordinate frames and rotations. This is important in the study of the motion of any type of vehicle such as airplanes, ships, and automobiles as well as mobile robots. It will be seen that frames provide an efficient means of keeping track of vehicle orientation and also enable simple conversion of displacement with respect to an intermediate frame to displacement with respect to a fixed frame.

3.2 Definition of Yaw, Pitch, and Roll

Shown in Figure 3.1 is a mobile robot with a coordinate frame attached. This frame moves with the robot and is called the robot frame. The y axis is aligned with the longitudinal axis of the robot, and the x axis points out the right side. The z axis points upward to form a right-handed system. This type of frame definition is commonly used in the field of robotics. It differs from the convention used by those in aerospace where the x axis is aligned with the longitudinal axis, the y axis is to the right, and the z axis points down, still a right-handed system.

As the robot moves about, it experiences translation or change in position. In addition to this, it may also experience rotation or change in attitude. The various rotations of the robot are now defined. Yaw is rotation about the z axis in the counter-clockwise direction as viewed looking into the z axis. Pitch is rotation about the new (after the yaw motion) x axis, in the counter-clockwise direction as viewed looking into the x axis, i.e., front end up is positive pitch. Roll is rotation about the new (after both yaw and pitch) y axis in the counter-clockwise direction as viewed looking into the y axis, i.e., left side of vehicle up is positive. In the system used by those in the aerospace field, pitch is

Mobile Robots: Navigation, Control and Sensing, Surface Robots and AUVs,
Second Edition. Gerald Cook and Feitian Zhang.
© 2020 by The Institute of Electrical and Electronics Engineers, Inc.
Published 2020 by John Wiley & Sons, Inc.

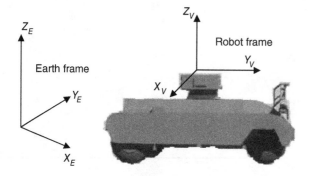

Figure 3.1 Mobile robot with earth and robot coordinate frames.

counter-clockwise rotation about the y axis while roll is counter-clockwise rotation about the x axis, i.e., the roles of the x and y axes are reversed with respect to these two rotations.

3.3 Rotation Matrix for Yaw

The rotation matrices for basic rotations are now derived. For yaw we have the diagram shown in Figure 3.2. Axes 1 represent the robot coordinate frame before rotation and axes 2 represent the robot coordinate frame after positive yaw rotation by the amount ψ. The z axes come out of the paper. It bears repeating that counter-clockwise rotation about the z axis is taken as positive yaw.

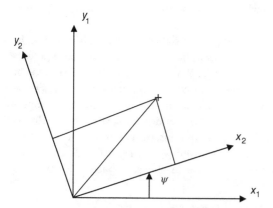

Figure 3.2 Frame 2 yawed with respect to frame 1.

We wish to express in the original coordinate frame 1 the location of a point whose coordinates are given in the new frame 2. For x and y we have

$$x_1 = x_2 \cos \psi - y_2 \sin \psi$$

$$y_1 = x_2 \sin \psi + y_2 \cos \psi$$

and for z

$$z_1 = z_2$$

or

$$\begin{bmatrix} x \\ y \\ z \end{bmatrix}_1 = \begin{bmatrix} \cos \psi & -\sin \psi & 0 \\ \sin \psi & \cos \psi & 0 \\ 0 & 0 & 1 \end{bmatrix} \begin{bmatrix} x \\ y \\ z \end{bmatrix}_2$$

Thus the rotation matrix for yaw is

$$R_{yaw}(\psi) = \begin{bmatrix} \cos \psi & -\sin \psi & 0 \\ \sin \psi & \cos \psi & 0 \\ 0 & 0 & 1 \end{bmatrix} \tag{3.1}$$

Example 1 *A vector expressed in the rotated coordinate system with ψ of $\pi/2$ is given by*

$$\begin{bmatrix} x \\ y \\ z \end{bmatrix}_2 = \begin{bmatrix} 1 \\ 0 \\ 0 \end{bmatrix}$$

Express this vector in the original coordinate system.

Solution 1

The expression of this vector in the original coordinate system becomes

$$\begin{bmatrix} x \\ y \\ z \end{bmatrix}_1 = \begin{bmatrix} \cos \pi/2 & -\sin \pi/2 & 0 \\ \sin \pi/2 & \cos \pi/2 & 0 \\ 0 & 0 & 1 \end{bmatrix} \begin{bmatrix} 1 \\ 0 \\ 0 \end{bmatrix} = \begin{bmatrix} 0 \\ 1 \\ 0 \end{bmatrix}$$

Note that the Euclidean norm of each column of the rotation matrix is one and that each column is orthogonal to each of the others. This is the definition of an orthonormal matrix. A convenient property of such matrices is that the inverse is simply the transpose, i.e.,

$$R_{yaw}(\psi)^{-1} = R_{yaw}(\psi)^{T} \tag{3.2}$$

or

$$R_{yaw}(\psi)^T R_{yaw}(\psi) = I$$

This property can be proved by premultiplying an orthonormal matrix by its transpose and then using the properties which it possesses, i.e.,

$$\langle col_i, col_j \rangle = 1, \quad i = j$$
$$= 0 \quad i \neq j$$

3.4 Rotation Matrix for Pitch

For pitch, we have the situation depicted in Figure 3.3. The x axes come out of the paper. Note again that front end up corresponds to positive pitch.

Again we wish to express in the original coordinate frame the location of a point whose coordinates have been given in the new frame. For x and z we have

$$y_1 = y_2 \cos \theta - z_2 \sin \theta$$
$$z_1 = y_2 \sin \theta + z_2 \cos \theta$$

and for x

$$x_1 = x_2$$

or

$$\begin{bmatrix} x \\ y \\ z \end{bmatrix}_1 = \begin{bmatrix} 1 & 0 & 0 \\ 0 & \cos \theta & -\sin \theta \\ 0 & \sin \theta & \cos \theta \end{bmatrix} \begin{bmatrix} x \\ y \\ z \end{bmatrix}_2$$

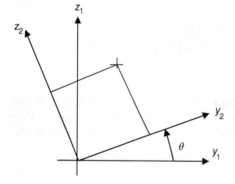

Figure 3.3 Frame 2 pitched with respect to frame 1.

Thus the rotation matrix for pitch is

$$R_{pitch}(\theta) = \begin{bmatrix} 1 & 0 & 0 \\ 0 & \cos\theta & -\sin\theta \\ 0 & \sin\theta & \cos\theta \end{bmatrix} \tag{3.3}$$

Example 2   *A vector expressed in the rotated coordinate system with θ of $\pi/2$ (i.e., pitched up by the angle $\pi/2$) is given by*

$$\begin{bmatrix} x \\ y \\ z \end{bmatrix}_2 = \begin{bmatrix} 0 \\ 1 \\ 0 \end{bmatrix}$$

Express this vector in the original coordinate system.

Solution 2

The expression of this vector in the original coordinate system becomes

$$\begin{bmatrix} x \\ y \\ z \end{bmatrix}_1 = \begin{bmatrix} 1 & 0 & 0 \\ 0 & \cos\pi/2 & -\sin\pi/2 \\ 0 & \sin\pi/2 & \cos\pi/2 \end{bmatrix} \begin{bmatrix} 0 \\ 1 \\ 0 \end{bmatrix} = \begin{bmatrix} 0 \\ 0 \\ 1 \end{bmatrix}$$

One may easily verify that the rotation matrix for pitch is also orthonormal.

3.5   Rotation Matrix for Roll

Finally we treat roll. This is counter-clockwise rotation about the y axis which results in left side up being defined as positive roll. The y axes come out of the paper as is shown in Figure 3.4.

Once more we wish to express in the original coordinate frame the location of a point whose coordinates are given in the new frame. For x and z we have

$$x_1 = x_2 \cos\phi + z_2 \sin\phi$$
$$z_1 = -x_2 \sin\phi + z_2 \cos\phi$$

and for y

$$y_1 = y_2$$

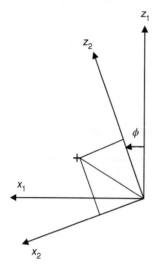

Figure 3.4 Frame 2 rolled with respect to frame 1.

or

$$
\begin{bmatrix} x \\ y \\ z \end{bmatrix}_1 = \begin{bmatrix} \cos\phi & 0 & \sin\phi \\ 0 & 1 & 0 \\ -\sin\phi & 0 & \cos\phi \end{bmatrix} \begin{bmatrix} x \\ y \\ z \end{bmatrix}_2
$$

Thus the rotation matrix for roll is

$$
R_{roll}(\phi) = \begin{bmatrix} \cos\phi & 0 & \sin\phi \\ 0 & 1 & 0 \\ -\sin\phi & 0 & \cos\phi \end{bmatrix} \tag{3.4}
$$

which is also orthonormal.

Example 3   *A vector expressed in the rotated coordinate system with ϕ of $\pi/2$ is given by*

$$
\begin{bmatrix} x \\ y \\ z \end{bmatrix}_2 = \begin{bmatrix} 1 \\ 0 \\ 0 \end{bmatrix}
$$

Express this vector in the original coordinate system.

Solution 3

In the original coordinate system the expression of this vector becomes

$$
\begin{bmatrix} x \\ y \\ z \end{bmatrix}_1 = \begin{bmatrix} \cos \pi/2 & 0 & \sin \pi/2 \\ 0 & 1 & 0 \\ -\sin \pi/2 & 0 & \cos \pi/2 \end{bmatrix} \begin{bmatrix} 1 \\ 0 \\ 0 \end{bmatrix} = \begin{bmatrix} 0 \\ 0 \\ -1 \end{bmatrix}
$$

Another way to think about the definitions of these different rotations is to reference them to the longitudinal axis of the vehicle, starting with the vehicle level and pointing along the y axis of the reference frame. Yaw is the rotation of the longitudinal axis of the robot in the horizontal plane. CCW rotation as viewed from above is taken as positive. Pitch is the rotation of the longitudinal axis of the robot in a plane perpendicular to the horizontal plane. Front end up is taken as positive. Roll is the rotation of the robot about its longitudinal axis. Left side up is taken as positive.

It is worth reiterating that each of these rotation matrices is orthonormal, i.e., the columns are all orthogonal to each other, and each column has Euclidean norm of one, making the inverse equal to the transpose.

3.6 General Rotation Matrix

We now define the general rotation matrix. After a frame has been yawed, pitched, and rolled, in this specific order, a point with coordinates given in this new frame may be converted into its coordinates in the original frame by the following operation

$$
\begin{bmatrix} x \\ y \\ z \end{bmatrix}_1 = R_{yaw}(\psi)R_{pitch}(\theta)R_{roll}(\phi) \begin{bmatrix} x \\ y \\ z \end{bmatrix}_2 \tag{3.5}
$$

Note that the conversion back into the original coordinates is in the reverse order of the rotations; i.e., roll was the last rotation. Therefore, it is the first matrix to operate on the coordinates of the point in question. Yaw was the first rotation; therefore, it is the last matrix to operate on the point in question. By multiplying these three rotation matrices together in the order shown above, we have the general rotation matrix:

$$
R(\psi,\theta,\phi) = \begin{bmatrix} \cos\psi\cos\phi - \sin\psi\sin\theta\sin\phi & -\sin\psi\cos\theta & \cos\psi\sin\phi + \sin\psi\sin\theta\cos\phi \\ \sin\psi\cos\phi + \cos\psi\sin\theta\sin\phi & \cos\psi\cos\theta & \sin\psi\sin\phi - \cos\psi\sin\theta\cos\phi \\ -\cos\theta\sin\phi & \sin\theta & \cos\theta\cos\phi \end{bmatrix}
\tag{3.6}
$$

It is easy to show that this product of orthonormal matrices is also orthonormal. Thus the general rotation matrix is also orthonormal.

As was the case for the individual rotation matrices, this general rotation matrix can be used to express a vector in an original coordinate frame when it has first been expressed in a frame that has been rotated with respect to the original frame. No matter what the attitude of a vehicle or how it arrived at this attitude, there exists a set of rotations in the order prescribed, yaw, pitch, and roll, which will yield this very same attitude.

A more generic expression of attitude that does not depend on one's choice of rotation order is the matrix comprised of direction cosines of the axes of frame 2 with the axes of frame 1. The components of the first column are successively the inner product of the x unit vector of frame 2 with the x unit vector of frame 1, the inner product of the x unit vector of frame 2 with the y unit vector of frame 1 and the inner produce of the x unit vector of frame 2 with the z unit vector of frame 1. Likewise, the components of the second column are successively the inner product of the y unit vector of frame 2 with the x unit vector of frame 1, the inner product of the y unit vector of frame 2 with the y unit vector of frame 1 and the inner produce of the y unit vector of frame 2 with the z unit vector of frame 1. Finally the components of the third column are successively the inner product of the z unit vector of frame 2 with the x unit vector of frame 1, the inner product of the z unit vector of frame 2 with the y unit vector of frame 1, and the inner produce of the z unit vector of frame 2 with the z unit vector of frame 1. In other words:

$$R_{21} = \begin{bmatrix} U_{x2}^T U_{x1} & U_{y2}^T U_{x1} & U_{z2}^T U_{x1} \\ U_{x2}^T U_{y1} & U_{y2}^T U_{y1} & U_{z2}^T U_{y1} \\ U_{x2}^T U_{z1} & U_{y2}^T U_{z1} & U_{z2}^T U_{z1} \end{bmatrix} \tag{3.7}$$

The entries of the matrix $R(\psi, \theta, \phi)$ given in Eq. (3.6) may be equated to this matrix yielding the values for yaw, pitch, and roll which when executed in that order would yield the given orientation. Equating terms it may be readily seen that

$$\tan \psi = -U_{y2}^T U_{x1} / U_{y2}^T U_{y1}$$

$$\sin \theta = U_{y2}^T U_{z1}$$

and

$$\tan \phi = -U_{x2}^T U_{z1} / U_{z2}^T U_{z1}$$

3.7 Homogeneous Transformation

There are situations where one frame is not only rotated with respect to another, but is also displaced. Suppose frame 2 is both rotated and displaced with respect to frame 1. Then a vector initially expressed with respect to frame 2 can be expressed with respect to frame 1 as below.

$$
\begin{bmatrix} x \\ y \\ z \end{bmatrix}_1 = R(\psi, \theta, \phi) \begin{bmatrix} x \\ y \\ z \end{bmatrix}_2 + \begin{bmatrix} x \\ y \\ z \end{bmatrix}_{\text{origin of frame 2 expressed in frame 1 coords}}
$$

or in shorthand notation

$$
\begin{bmatrix} x \\ y \\ z \end{bmatrix}_1 = R_{21} \begin{bmatrix} x \\ y \\ z \end{bmatrix}_2 + \begin{bmatrix} x_o \\ y_o \\ z_o \end{bmatrix}_{21} \tag{3.8}
$$

If one goes through a series of transformations, the operations become even more cumbersome. For the case of two transformations the equations are

$$
\begin{bmatrix} x \\ y \\ z \end{bmatrix}_2 = R_{32} \begin{bmatrix} x \\ y \\ z \end{bmatrix}_3 + \begin{bmatrix} x_o \\ y_o \\ z_o \end{bmatrix}_{32}
$$

and

$$
\begin{bmatrix} x \\ y \\ z \end{bmatrix}_1 = R_{21} \begin{bmatrix} x \\ y \\ z \end{bmatrix}_2 + \begin{bmatrix} x_o \\ y_o \\ z_o \end{bmatrix}_{21}
$$

or

$$
\begin{bmatrix} x \\ y \\ z \end{bmatrix}_1 = R_{21}R_{32} \begin{bmatrix} x \\ y \\ z \end{bmatrix}_3 + R_{21}\begin{bmatrix} x_o \\ y_o \\ z_o \end{bmatrix}_{32} + \begin{bmatrix} x_o \\ y_o \\ z_o \end{bmatrix}_{21}
$$

This can be written more concisely as a single operation using the *homogeneous transformation*. For a single transformation containing translation and rotation

$$
\begin{bmatrix} x \\ y \\ z \\ 1 \end{bmatrix}_1 = A_{21} \begin{bmatrix} x \\ y \\ z \\ 1 \end{bmatrix}_2 \tag{3.9}
$$

where for A_{21} we have

$$
A_{21} = \begin{bmatrix} & & & x_o \\ & R_{21} & & y_o \\ & & & z_o \\ 0 & 0 & 0 & 1 \end{bmatrix} \tag{3.10}
$$

Note that the upper left three-by-three matrix is the rotation matrix while the upper portion of the right column is comprised of the origin of frame 2 in frame 1 coordinates. Here x_o, y_o, and z_o could represent, for example, the origin of the sensor frame in vehicle coordinates. One can use this transformation to convert a vector specified in one set of coordinates to its expression in another set of coordinates in a single operation. When using this homogeneous transformation, the position vectors are converted to dimension four by appending a 1 as the fourth entry. This is necessary not only to make the matrix operations conformal, but also to couple in the location of the origin of the second coordinate system with respect to the original coordinate system.

Example 4 *Let frame 2 be both rotated and displaced with respect to frame 1. The rotation is a yaw of 90°*

$$R_{21} = \begin{bmatrix} 0 & -1 & 0 \\ 1 & 0 & 0 \\ 0 & 0 & 1 \end{bmatrix}$$

and the displacement of the origin of frame 2 with respect to frame 1 is

$$\begin{bmatrix} x_0 \\ y_0 \\ z_0 \end{bmatrix} = \begin{bmatrix} 10 \\ 5 \\ 0 \end{bmatrix}$$

Now let the point of interest be given by

$$\begin{bmatrix} x \\ y \\ z \end{bmatrix}_{expressed\ in\ frame\ 2} = \begin{bmatrix} 1 \\ 0 \\ 0 \end{bmatrix}$$

Express this vector in frame 1.

Solution 4

In frame 1 the expression of this vector becomes

$$\begin{bmatrix} x \\ y \\ z \end{bmatrix}_{expressed\ in\ frame\ 1} = \begin{bmatrix} 0 & -1 & 0 \\ 1 & 0 & 0 \\ 0 & 0 & 1 \end{bmatrix} \begin{bmatrix} 1 \\ 0 \\ 0 \end{bmatrix} + \begin{bmatrix} 10 \\ 5 \\ 0 \end{bmatrix}$$

or

$$\begin{bmatrix} x \\ y \\ z \end{bmatrix}_{expressed\ in\ frame\ 1} = \begin{bmatrix} 0 \\ 1 \\ 0 \end{bmatrix} + \begin{bmatrix} 10 \\ 5 \\ 0 \end{bmatrix} = \begin{bmatrix} 10 \\ 6 \\ 0 \end{bmatrix}$$

Now solving this problem by using the homogeneous transformation matrix we have

$$\begin{bmatrix} x \\ y \\ z \\ 1 \end{bmatrix}_{\text{expressed in frame 1}} = \begin{bmatrix} 0 & -1 & 0 & 10 \\ 1 & 0 & 0 & 5 \\ 0 & 0 & 1 & 0 \\ 0 & 0 & 0 & 1 \end{bmatrix} \begin{bmatrix} 1 \\ 0 \\ 0 \\ 1 \end{bmatrix}$$

or

$$\begin{bmatrix} x \\ y \\ z \\ 1 \end{bmatrix}_{\text{expressed in frame 1}} = \begin{bmatrix} 10 \\ 6 \\ 0 \\ 1 \end{bmatrix}$$

The homogeneous transformation matrices can be multiplied just as the rotation matrices can. Thus the homogeneous transformation to take a vector from its expression in frame 3 coordinates to its expression in frame 2 coordinates and finally to its expression in frame 1 coordinates can be written

$$\begin{bmatrix} x \\ y \\ z \\ 1 \end{bmatrix}_1 = [A_{21}][A_{32}] \begin{bmatrix} x \\ y \\ z \\ 1 \end{bmatrix}_3 \tag{3.11}$$

or

$$\begin{bmatrix} x \\ y \\ z \\ 1 \end{bmatrix}_1 = [A_{31}] \begin{bmatrix} x \\ y \\ z \\ 1 \end{bmatrix}_3$$

where

$$[A_{31}] = [A_{21}][A_{32}]$$

Another interesting and useful property of the homogeneous transformation matrix is that its inverse can be expressed as

$$\begin{bmatrix} & & & x_o \\ & R_{21} & & y_o \\ & & & z_o \\ 0 & 0 & 0 & 1 \end{bmatrix}^{-1} = \begin{bmatrix} & & & a \\ & R_{21}^T & & b \\ & & & c \\ 0 & 0 & 0 & 1 \end{bmatrix} \tag{3.12}$$

where the entries in the upper part of the last column are defined by

$$\begin{bmatrix} a \\ b \\ c \end{bmatrix} = -R_{21}^T \begin{bmatrix} x_o \\ y_0 \\ z_o \end{bmatrix} \tag{3.13}$$

In all of these, use has been made of the fact that for a rotation matrix

$$R^{-1} = R^T$$

since the rotation matrix is orthonormal.

This homogeneous transformation provides a concise means of expressing a vector in an original frame when the second frame is both rotated and translated with respect to the original frame. Its convenience becomes even more pronounced when there is a series of transformations, e.g., sensor frame to vehicle frame and then vehicle frame to earth frame.

3.8 Rotating a Vector

Another important application of the rotation matrix is to express the new coordinates of a vector after the vector itself has been yawed, pitched, and rolled. Here the same coordinate frame is used before and after the rotation. To illustrate, consider an initial vector expressed in frame 1. This vector is now rotated about the z axis. The expression for this rotated vector, again in frame 1, is given by the following

$$\begin{bmatrix} x \\ y \\ z \end{bmatrix}_{\text{after rotation}} = \begin{bmatrix} \cos\psi & -\sin\psi & 0 \\ \sin\psi & \cos\psi & 0 \\ 0 & 0 & 1 \end{bmatrix} \begin{bmatrix} x \\ y \\ z \end{bmatrix}_{\text{before rotation}}$$

or

$$\begin{bmatrix} x \\ y \\ z \end{bmatrix}_{\text{after rotation}} = R_{yaw}(\psi) \begin{bmatrix} x \\ y \\ z \end{bmatrix}_{\text{before rotation}} \tag{3.14}$$

This same process holds for each of the rotations. Thus, if a vector is rotated first about the y axis, then about the x axis, and finally about the z axis, the new vector in the original frame is given by

$$\begin{bmatrix} x \\ y \\ z \end{bmatrix}_{\text{after rot}} = \begin{bmatrix} \cos\psi\cos\phi - \sin\psi\sin\theta\sin\phi & -\sin\psi\cos\theta & \cos\psi\sin\phi + \sin\psi\sin\theta\cos\phi \\ \sin\psi\cos\phi + \cos\psi\sin\theta\sin\phi & \cos\psi\cos\theta & \sin\psi\sin\phi - \cos\psi\sin\theta\cos\phi \\ -\cos\theta\sin\phi & \sin\theta & \cos\theta\cos\phi \end{bmatrix} \begin{bmatrix} x \\ y \\ z \end{bmatrix}_{\text{before rot}}$$

or

$$\begin{bmatrix} x \\ y \\ z \end{bmatrix}_{after\ rotation} = R(\psi, \theta, \phi) \begin{bmatrix} x \\ y \\ z \end{bmatrix}_{before\ rotation} \tag{3.15}$$

To reiterate, in this second application of rotation matrices, the vector on the right-hand side of the equation is the vector before its rotation and the result on the left-hand side is the vector after its rotation. Both vectors are expressed in the same frame.

We shall find important uses for these rotation matrices and homogeneous transformation matrices in the chapters that follow.

Exercises

1   Evaluate the rotation matrix for the case where $\psi = \pi/2$, $\theta = -\pi/2$, and $\phi = 0$.

2   Evaluate the rotation matrix for the case where $\psi = 0$, $\theta = \pi/2$, and $\phi = \pi/2$.

3   Evaluate the homogeneous transformation for the case where the second frame has orientation with respect to the first frame of $\psi = \pi/2$, $\theta = -\pi/2$, and $\phi = 0$ and location with respect to the first frame of $x = 3$, $y = 2$, and $z = 1$.

4   Evaluate the homogeneous transformation for the case where the second frame has orientation with respect to the first frame of $\psi = 0$, $\theta = \pi$, and $\phi = \pi/2$ and location with respect to the first frame of $x = 1$, $y = 3$, and $z = 2$.

5   Given a target whose location is expressed in frame 2 as $x = 1$, $y = 2$, and $z = 0$, find its location with respect to frame 1. The origin of frame 2 is at $x = 20$, $y = 10$, and $z = 1$ with respect to frame 1. The orientation of frame 2 with respect to frame 1 is $\psi = \pi/2$, $\theta = \pi/4$, and $\phi = 0$. Solve using a rotation plus a translation and also solve using the homogeneous transformation.

6   A target is located in frame 3 at coordinates $x = 3$, $y = 2$, and $z = 1$. The origin of frame 3 is at coordinates $x = 10$, $y = 0$, and $z = 0$ with respect to frame 2. The orientation of frame 3 with respect to frame 2 is $\psi = \pi/4$, $\theta = -\pi/4$, and $\phi = 0$. The origin of frame 2 is at coordinates $x = 0$, $y = 5$, and $z = 0$ with respect to frame 1. The orientation of frame 2 with respect to frame 1 is $\psi = \pi$, $\theta = 0$, and $\phi = \pi/2$.

A Compute the homogeneous transformation describing frame 3 with respect to frame 2 and determine the location of the target in frame 2.

B Next compute the homogeneous transformation describing frame 2 with respect to frame 1 and determine the location of the target in frame 1.

C Finally multiply the homogeneous transformations together (in the proper order) and determine the location of the target in frame 1 in a single step.

A vector has coordinates $[1 \ 0 \ 0]'$. This vector is to be rotated about the y axis by the amount $\pi/2$. Use the rotation matrix to determine the resulting vector.

7 A vector has coordinates $[1 \ 0 \ 0]'$. This vector is to be rotated about the y axis by the amount $\pi/2$ and then about the z axis by the amount $-\pi/2$. Use the rotation matrix to determine the resulting vector.

References

Brogan, W. L., *Modern Control Theory*, Prentice Hall, Upper Saddle River, NJ, 1991.

Kuipers, J. B., *Quaternions and Rotation Sequences*, Princeton University Press, Princeton, NJ, 1999.

Lavalle, S. M., *Planning Algorithms/Motion Planning*, Cambridge University Press, 2006.

4

Robot Navigation

4.1 Introduction

This chapter introduces the topic of navigation and the various means of accomplishing this. The focus is on Inertial Navigation Systems (INS) (gimbaled and strap-down) and the Global Positioning System (GPS). Also, briefly discussed is deduced reckoning utilizing less sophisticated methodology.

4.2 Coordinate Systems

One definition of navigation is the process of accurately determining position and velocity relative to a known reference or the process of planning and executing the maneuvers necessary to move between desired locations. One important factor in navigation is an understanding of the different coordinate systems. Figure 4.1 shows a sphere representing the earth along with several coordinate frames. To minimize confusion, only the x and z axes are shown. The y axes in each case are such as to form right-handed coordinate systems.

4.3 Earth-Centered Earth-Fixed Coordinate System

In Coordinate System I, the z axis points out the North pole, the x axis points through the equator at the prime meridian, and the y axis (not shown) completes the right-handed coordinate frame. This set of axes is called the earth-centered earth-fixed axes (ECEF). As the name implies, this set of axes has its origin at the center of the earth and rotates with the earth. There is a unique relation between the ECEF coordinates of a point on the surface of the earth and

Mobile Robots: Navigation, Control and Sensing, Surface Robots and AUVs,
Second Edition. Gerald Cook and Feitian Zhang.
© 2020 by The Institute of Electrical and Electronics Engineers, Inc.
Published 2020 by John Wiley & Sons, Inc.

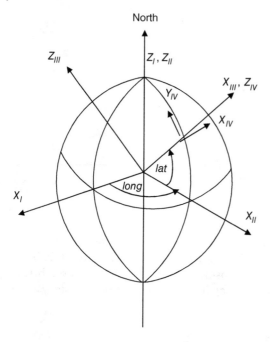

North

Figure 4.1 Earth and several different coordinate frames.

its longitude, which is measured positively Eastward from the prime meridian running through Greenwich, England, and its latitude, which is measured positively Northward from the equator. Starting with latitude and longitude the X, Y, and Z in ECEF coordinates can be determined approximately assuming a spherical model of the earth of radius R and using the equations

$$X = R \cos(lat) \cos(long) \tag{4.1a}$$

$$Y = R \cos(lat) \sin(long) \tag{4.1b}$$

$$Z = R \sin(lat) \tag{4.1c}$$

It should be pointed out that the earth is not a perfect sphere and that more precise models of its shape do exist. These more precise models account for the flatness of the earth, i.e., the fact that the radius at the poles, 6,356.7 km, is slightly less than the radius at the equator, 6,378.1 km. The spherical model is used in examples here for its simplicity in application.

Example 1 *For long = 85°W and lat = 42°N find X, Y, and Z in ECEF coordinates. Use dimensions of meters and assume the point is on the earth's surface.*

Solution 1

Since the longitude is 85° to the West, it is expressed as −85°.

$$X = R\ cos\ (lat)\ cos\ (long) = 6,378,137\ cos\ (42)\ cos\ (-85)$$
$$= 413,107.719\ m$$

$$Y = R\ cos\ (lat)\ sin\ (long) = 6,378,137\ cos\ (42)\ sin\ (-85)$$
$$= -4,721,842.835\ m$$

$$Z = R\ sin\ (lat) = 6,378,137\ sin\ (42) = 4,267,806.678\ m$$

In the example above, we have used the equatorial radius of the earth for R. The process can be reversed if the ECEF coordinates are given and the latitude and longitude have to be determined.

$$lat = \tan^{-1}\left(\frac{Z}{\sqrt{X^2 + Y^2}}\right) \tag{4.2a}$$

$$long = \tan^{-1}\left(\frac{Y}{X}\right) \tag{4.2b}$$

Example 2 For the point at ECEF coordinates, $X = 3,000,000\ m$, $Y = -5,000,000\ m$, and $Z = -2,638,181\ m$, find lat and long (Note: The point may be slightly off the earth's surface.).

Solution 2

$$X = 3,000,000$$

$$Y = -5,000,000$$

$$Z = -2,638,181$$

$$Long = (180/\pi) \times \tan^{-1}\left(\frac{-5 \times 10^6}{3 \times 10^6}\right) = -59.03° \text{ or } 59.03° W$$

$$\sqrt{X^2 + Y^2} = 5.83095 \times 10^6$$

$$Lat = (180/\pi) \times \tan^{-1}\left(\frac{-2.638181 \times 10^6}{5.83095 \times 10^6}\right) = -24.4° \text{ or } 24.4° S$$

$$R = \sqrt{X^2 + Y^2 + Z^2} = 6,400,000\ m$$

4.4 Associated Coordinate Systems

Other coordinates are useful in describing motion on the surface of the earth, and some of these shown in Figure 4.1 are now described. The relationships between the variables of coordinate systems II and I are given below. Here, the angles referred to as *lat* and *long* are assumed to be expressed in radians while angles referred to as *lat* and *long* are assumed to be expressed in degrees.

$$
\begin{bmatrix} X \\ Y \\ Z \end{bmatrix}_{II} = \begin{bmatrix} \cos long & \sin long & 0 \\ -\sin long & \cos long & 0 \\ 0 & 0 & 1 \end{bmatrix} \begin{bmatrix} X \\ Y \\ Z \end{bmatrix}_{I} \tag{4.3}
$$

Coordinate frame *II* has been rotated counter clockwise about the Z_I axis by an amount *long*. This corresponds to a new frame with the X_{II} axis now pointing through the equator at longitude *long*. Note that this matrix is given by $R_{yaw}^{-1}(long)$ or $R_{yaw}^{T}(long)$.

The relationships between the variables of coordinate systems III and II are given below.

$$
\begin{bmatrix} X \\ Y \\ Z \end{bmatrix}_{III} = \begin{bmatrix} \cos lat & 0 & \sin lat \\ 0 & 1 & 0 \\ -\sin lat & 0 & \cos lat \end{bmatrix} \begin{bmatrix} X \\ Y \\ Z \end{bmatrix}_{II} \tag{4.4}
$$

Coordinate frame *III* has been rotated clockwise about the Y_{II} axis by an amount *lat*. The X_{III} axis now points through the meridian of longitude *long* and the parallel of latitude *lat*. Note that this matrix is given by $R_{roll}^{-1}(-lat)$ or $R_{roll}^{T}(-lat)$ or $R_{roll}(lat)$.

The relationships between the variables of coordinate systems *IV* and *III* are given below.

$$
\begin{bmatrix} X \\ Y \\ Z \end{bmatrix}_{IV} = \begin{bmatrix} 0 & 1 & 0 \\ 0 & 0 & 1 \\ 1 & 0 & 0 \end{bmatrix} \begin{bmatrix} X \\ Y \\ Z \end{bmatrix}_{III} + \begin{bmatrix} 0 \\ 0 \\ -R \end{bmatrix} \tag{4.5}
$$

For coordinate frame *IV*, the origin has been moved from the center of the earth to the surface of the earth. The Y_{IV} axis is parallel to the Z_{III} axis, the Z_{IV} axis is parallel to the X_{III} axis, and the X_{IV} axis is parallel to the Y_{III} axis. One can think of the orientation of frame *IV* as one obtained by rotating frame *III* about its z axis by 90° ccw and then rotating about the new x axis by 90° ccw. The rotation matrix is $[R_{yaw}(\pi/2)R_{pitch}(\pi/2)]^{-1}$ or $[R_{yaw}(\pi/2)R_{pitch}(\pi/2)]^{T}$ where

$$R_{yaw}(\pi/2) = \begin{bmatrix} 0 & -1 & 0 \\ 1 & 0 & 0 \\ 0 & 0 & 1 \end{bmatrix}$$

and

$$R_{pitch}(\pi/2) = \begin{bmatrix} 1 & 0 & 0 \\ 0 & 0 & -1 \\ 0 & 1 & 0 \end{bmatrix}$$

This coordinate frame attached to the surface of the earth with the y axis pointing North, the X_{IV} axis pointing East, and the Z_{IV} axis pointing outward from the earth's surface is a useful local coordinate system. One can describe x–y locations with respect to this frame in terms of longitude and latitude given the longitude and latitude of the origin of the coordinate system.

By assuming a spherical earth and performing the above operations in succession, starting with an initial point

$$\begin{bmatrix} X \\ Y \\ Z \end{bmatrix}_I = \begin{bmatrix} R \cos long \cos lat \\ R \sin long \cos lat \\ R \sin lat \end{bmatrix}$$

and defining the latitude and longitude of the origin of the final frame to be $long_0$ and lat_0 we get:

$$X_{IV} = -R \cos long \sin long_0 \cos lat + R \sin long \cos long_0 \cos lat$$
$$Y_{IV} = -R \cos long \cos long_0 \cos lat \sin lat_0$$
$$\qquad - R \sin long \sin long_0 \cos lat \sin lat_0 + R \sin lat \cos lat_0$$
$$Z_{IV} = R \cos (long_0 - long) \cos lat \cos lat_0 + R \sin lat_0 \sin lat - R$$

which reduce to

$$X_{IV} = R \cos lat \sin (long - long_0) \tag{4.6a}$$
$$Y_{IV} = -R \cos lat \sin lat_0 \cos (long - long_0) + R \sin lat \cos lat_0$$
$$\tag{4.6b}$$

and

$$Z_{IV} \approx R \cos lat \cos lat_0 + R \sin lat_0 \sin lat - R = R \cos (lat - lat_0) - R$$
$$\tag{4.6c}$$

For points on the surface of the earth in the vicinity of the origin of the final frame, these equations for x and y may be approximated quite accurately as

$$X_{IV} \approx R \cos (lat) [long - long_0] \tag{4.7a}$$

$$Y_{IV} \approx R[lat - lat_0] \tag{4.7b}$$

and

$$Z_{IV} \approx 0 \tag{4.7c}$$

Example 3 *A local coordinate system is set up at long = $70°W$ = $-70°$ and lat = $38°N$. A mobile robot is at long = $69.998°W$ = $-69.998°$ and lat = 38.001 $°N$. Find the X, Y coordinates for the robot. Take X-East and Y-North.*

Solution 3

$$\begin{aligned}
X_{local} &= R\left(long - long_0\right) \cos\left(lat_0\right)(\pi/180) \\
&= 6,378,137(0.002)(0.788)(\pi/180) \\
&= 175.4\ m
\end{aligned}$$

$$Y_{local} = R(lat - lat_0)(\pi/180) = 6,378,137(0.001)(\pi/180) = 111.3\ m$$

If we now go to a final local coordinate system rotated such that the x axis of frame V is at an angle α with respect to the x axis of frame IV, i.e., East, then we have the frame shown in Figure 4.2.

The appropriate rotation matrix is given by

$$\begin{bmatrix} X \\ Y \\ Z \end{bmatrix}_V = \begin{bmatrix} \cos\alpha & \sin\alpha & 0 \\ -\sin\alpha & \cos\alpha & 0 \\ 0 & 0 & 1 \end{bmatrix} \begin{bmatrix} X \\ Y \\ Z \end{bmatrix}_{IV} \tag{4.8}$$

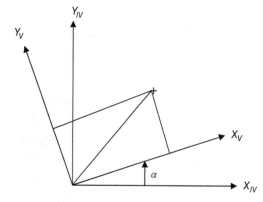

Figure 4.2 Local coordinate system with X axis rotated relative to east.

Note that this transformation matrix is given by $R_{yaw}(\alpha)^{-1}$ or $R_{yaw}(\alpha)^{T}$. For all of these transformations in the preceding, the inverse (or transpose) is required because the coordinates are being converted from their expression in the old frame to their expression in the new frame rather than vice versa as was the case considered as the rotation matrices were derived.

Applying the above transformation to equation (4.7) yields as the expression for the coordinates in frame V

$$X_V = R \cos \alpha \cos lat_0 [long - long_0] + R \sin \alpha [lat - lat_0] \tag{4.9a}$$

and

$$Y_V = - R \sin \alpha \cos lat_0 [long - long_0] + R \cos \alpha [lat - lat_0] \tag{4.9b}$$

Example 4 *Re-do the previous example but with X–Y axes now rotated by 30° clockwise.*

Solution 4

$$X_{local} = R \cos (\alpha)(long - long_0) \cos (lat_0)(\pi/180)$$
$$+ R \sin (\alpha)(lat - lat_0)(\pi/180)$$
$$X_{local} = 175.4 \cos (-30) + 111.3 \sin (-30) = 96.24 \, m$$
$$Y_{local} = - R \sin (\alpha)(long - long_0) \cos (lat_0)(\pi/180)$$
$$+ R \cos (\alpha)(lat - lat_0)(\pi/180)$$
$$Y_{local} = - 175.4 \sin (-30) + 111.3 \cos (-30) = 184.09 \, m$$

4.5 Universal Transverse Mercator Coordinate System

A commonly used coordinate system is the Universal Transverse Mercator (UTM) Coordinate System. The more common Mercator projections result from projecting the sphere onto a cylinder tangent to the equator. Regions near the poles are greatly distorted appearing larger than they really are. Regions near the equator are most accurate. The purpose of such a projection is to convert the spherical shape of the earth to a flat map.

Transverse Mercator Projections result from projecting the sphere onto a cylinder tangent to a central meridian. Regions near the central meridian are projected most accurately. Distortion of scale, distance, direction, and area increase as one moves away from the central meridian. In the UTM coordinate system, longitudinal zones are only 6° of longitude wide, extending 3° to either side of the central meridian. Transverse Mercator maps are often used to portray areas with larger north–south than east–west extent. In the UTM coordinate system, these 6° longitudinal zones extend from 80 °S latitude to 84 °N latitude. There are 60 of these longitudinal zones covering the entire earth labeled with the numbers 1–60. Each longitudinal zone is further divided into zones of latitude, beginning with zone C at 80 °S up to M just below the equator. To the North, the zones run from N just above the equator to X at 84 °N. All the zones span 8° in the north–south direction except zone X, which spans 12°.

Within each longitudinal zone, the easting coordinate is measured from the central meridian with 500 km false easting added to ensure positive coordinates. The northing coordinate is measured from the equator with a 10,000 km false northing added for positions south of the equator. The coordinates thus derived define a location within a UTM longitudinal zone either north or south of the equator, but because the same co-ordinate system is repeated for each zone and hemisphere, it is necessary to not only state the northing and easting but to also state the UTM longitudinal zone and either the hemisphere or latitudinal zone to define the location uniquely world-wide.

The following are formulas relating latitude and longitude to UTM. Given a latitude and longitude, the UTM coordinates can be determined by first computing the longitudinal zone number

$$i = int\left(\frac{180 + long}{6}\right) + 1 \tag{4.10}$$

Here longitude in the westward direction would be taken as negative. The Central Meridian for the longitudinal zone is then given by:

$$Long_0 = [-177 + (i-1)6] \tag{4.11}$$

Using the spherical earth approximation and also ignoring the distortion in projecting the sphere onto the cylinder, one may roughly compute northern and eastern as

$$Northing \approx R\,lat\,\pi/180 \tag{4.12a}$$

$$Easting \approx R(long - long_0)(\pi/180)\cos(lat) + 500,000 \tag{4.12b}$$

The correct calculation of northing and easting uses ellipsoids to account for the true shape of the earth and the distortion caused by the projection. Among other effects, this model yields a larger northern for points lying off the central meridian than for points of the same latitude lying on the central meridian.

Example 5 *Find the UTM coordinates for the point long = 10°E, lat = 43°N.*

Solution 5

Again a spherical model for the earth is used with R = 6, 378, 137 m. Also, the distortion in projecting the strip about the central meridian onto the cylinder is ignored. Using these approximations yields the following rough values for the solution

$$Long = 10°E, \quad lat = 43°N$$

$$i = int\left(\frac{180 + long}{6}\right) + 1$$

$$i = 32$$

Central Meridian Calculation

$$Long_0 = [-177 + (32 - 1)6] = 9°$$

$$Long - long_0 = 10 - 9 = 1°$$

$$Northing \approx R \, lat(\pi/180) = 4,786,938.104 \, m$$

$$Easting \approx R(long - long_0)(\pi/180) \, cos\,(43°) + 500,000 = 581,413.92 \, m$$

The UTM system is sometimes preferable to longitude and latitude in specifying relative locations because of the linear scale and because most persons are familiar with the unit of meters.

4.6 Global Positioning System

The GPS provides a means for a receiver or user to determine its location anywhere on or slightly above the surface of the earth. This is sometimes referred to as geolocation. The GPS system includes a constellation of 24 earth-orbiting satellites which are situated in such a way as to maximize coverage of the earth. Their orbital radii are approximately 20,200 km, and they are spaced in six orbits with four satellites per orbit. The orbits have inclination angles of 55° with respect to the equator, and their orbital period is 12 hours. Each satellite is equipped with an atomic clock and a radio transmitter and receiver.

The status and operational capability of the satellites is monitored by a series of ground stations with antennas stationed in different parts of the world as well as a master control station. The entire operation depends on the use of encoded radio signals. The Standard Positioning Service utilizes a 1.023 MHz repeating pseudo-random code called Coarse Acquisition (C/A) code and is available for public use. The Precise Positioning Service utilizes a 10.23 MHz repeating pseudo-random code called Precise Acquisition (P) code. It can be encrypted to make it available for use by the Department of Defense only.

Geolocation is based on the use of modulated radio signals transmitted from the satellites and received by the user. Based on signal travel time one can determine distance. Distance calculations from the user to the visible satellites combined with knowledge of the satellite positions at the time of the signal transmission allow one to use triangulation and thereby determine the user location.

By computing the time duration of one bit in the pseudo-random code and multiplying this by the speed of light, one can determine the potential resolution of distance calculations from the satellite to the user. The C/A Code provides the potential for distance resolution of 30 m or better while the P code provides the potential for distance resolution of 3 m or better. GPS systems used in the surveying mode, where the receiver is stationary for hours, have the capability for distance resolution in the centimeter range.

The GPS ground stations as well as the GPS satellites utilize atomic clocks. Time is measured starting at $24:00:00$, January 5, 1980. No leap seconds are included in GPS Time. Receiver clocks are not as accurate as the atomic clocks and normally exhibit bias. This bias creates errors in the determination of travel time of the signals and therefore causes errors in calculation of the distances to the satellites. These distances to all satellites will be in error by the amount of the local clock error multiplied by the speed of light.

In addition to distance errors caused by the local clock error, atmospheric effects can also cause errors. There are ionospheric delays caused by the layer of the atmosphere containing ionized air, and there are also tropospheric delays caused by changes in the temperature, pressure, and humidity of the lowest part of the atmosphere.

Apart from geolocation errors caused by errors in the distance calculations to the satellites, there are also geolocation errors caused by incorrect satellite ephemeris data. The ephemeris errors may be decomposed into tangential, radial, and cross track components. Radial ephemeris errors have the greatest impact on geolocation errors.

Another source of error is caused by multipath transmission. Here reflected signals near the receiver may interfere or be mistaken for the original signal. Because multipath signals have a longer route, the computed distance from the satellite will be greater than the actual distance. Multipath transmission is difficult to detect and sometimes hard to avoid.

GPS receivers receive and process signals from the in-view satellites. In the case of the P code which provides most precise geolocation, a unique segment of the 10.23 MHz PRN code is generated at each satellite and is known ahead of time by the receiver. This P code operating at 10.23 chips per microsecond repeats only once per week. It is combined with 50 bps data sequences via the exclusive-or operation. The data sequence, which consists of time-tagged data bits marking the time they are transmitted, is sent every 30 seconds and contains the Navigation Message, which is information regarding the GPS satellite orbits, clock correction, and other system parameters. The carrier signal is either the L1 sinusoid operating at 1,575.42 MHz or the L2 sinusoid operating at 1,227.6 MHz modulated by the data-modulated 10.23 MHz PRN using binary phase shift keying.

In the case of the C/A code which provides less precise geolocation, a 1.023 MHz PRN code is generated at each satellite, and it is also known ahead of time by the receiver. This C/A code operating at 1.023 chips per microsecond repeats once per millisecond. It too is combined with 50 bps data sequences via the exclusive-or operation. The data sequence is sent every 30 seconds and contains the Navigation Message. The carrier signal here is the L1 sinusoid operating at 1,575.42 MHz.

Basically, the system operates by the receiver noting on its local clock the time at which signals are received from satellites. This time of arrival is computed by shifting within the receiver the known segment of the P code for the particular satellite and correlating it with the received signal. The correlation will be maximized when the shift corresponds to the time of arrival. Correlators have been developed that permit simultaneous correlations of the received signal with thousands of different time shifted signals providing rapid determination of the time of arrival. Under normal operation the 50 bps data sequence is not known beforehand. This limits the duration of the time sequences used for the correlation to 1/50th of a second. Usually this duration of signal is sufficient, but under noisy environments this limitation can cause a problem. Longer correlation durations have the effect of reducing the noise impact inversely to correlation duration, thereby making the determination of the correlation peak more accurate.

Knowing the time the signal was transmitted from the time-tagged data, and having determined the time of arrival with respect to the receiver clock, i.e., the local clock, the travel time for each received signal may be computed and converted to the distance from the receiver to the respective satellites. This travel time will be accurate within local clock error plus the error in correlator alignment. If correlator alignment is correct within one chip of the P code, this corresponds to $(3 \times 10^8 \text{ m/s})/(10.23 \times 10^6 \text{ Hz})$ or approximately 30 m. If the correlator alignment is correct within one hundredth of a chip width, the error in distance would be less than 0.3 m. These figures assume that the local clock error has been perfectly accounted for. In the case of the C/A code, the error in distance caused by an alignment error of one chip is approximately 300 m. If the correlator alignment is correct within one hundredth of a chip width, the error in distance would be less than 3 m. The result of performing these correlations on signals received from all the visible satellites is a set of pseudo-ranges from the receiver to these visible GPS satellites.

Knowing the distance to a single satellite, along with the knowledge of that satellite's position at the time the signal was transmitted, places the receiver on a sphere centered about that point. With signals from two satellites, the receiver is placed on a sphere about each of two points with their intersection being a circle. Using a measurement from a third satellite places the receiver on a sphere about this third point. The intersection of this sphere with the previously described circle yields two points, and only one of these is near the surface of the earth. Thus, in principle, three GPS satellites are sufficient to locate the receiver if there is no local clock error. In practice, an error does exist between the local clock and the GPS clock of the fleet of satellites. In order to correct for this local clock error, signals must be received from a fourth satellite. This extra equation allows one to determine the three-dimensional position as well as the local clock error. If more than four satellites are visible, the redundancy can be used to reduce other types of errors.

Geometric Dilution of Precision (GDOP) is computed from the geometric relationships between the receiver position and the positions of the satellites the receiver is using for navigation. If there is not a good spread among the visible satellites, GDOP will be high and the computed position of the receiver is more sensitive to small errors in distance calculations and satellite positions. Imagine that two satellites are close together. The distance from each of these satellites to the receiver yields a sphere about the respective satellite. Since the two spheres are approximately the same size and have approximately the same center, their intersection will be very sensitive to any kind of error. GDOP components include position dilution of precision, horizontal dilution of precision, vertical dilution of precision, and time dilution of precision.

Differential GPS (DGPS) provides improved precision in the computed location of the receiver. Here one receiver, a base station, is placed at a surveyed location, and the other receivers are free to rove. The difference between the computed position and the known true position for the base station is evaluated. These errors and information about the different factors contributing to the errors are broadcast to all the roving receivers for their use. By this means the GPS accuracy can be substantially improved through canceling the effect of the common-mode errors. The effectiveness of DGPS degrades when the rovers are separated from the base station by as much as tens of miles. For this system to be successful, the base station must, at a minimum, broadcast the following set of information for each satellite:

- Satellite identification number
- Range correction
- Ephemeris set identifier
- Reference time

4.7 Computing Receiver Location Using GPS, Numerical Methods

Having performed the correlations of the signals received from the visible satellites with the shifted signals generated within the receiver, the times of arrival of the signals from the satellites may be extracted. Then the travel times are determined and the pseudo-distances from the receiver to the visible satellites are computed. Once this has been accomplished, one can proceed to an iterative process for the determination of the receiver location.

4.7.1 Computing Receiver Location Using GPS via Newton's Method

The following is the system of nonlinear equations that are based on the measurements of distances from four or more different satellites to the receiver. Here the d^i are computed from travel time of the signals multiplied by the speed of light, (x^i, y^i, z^i) are the ECEF coordinates of satellite i, (x, y, z) are the assumed ECEF position coordinates of the receiver antenna, t_b is the receiver clock bias, and c is the speed of light. The unknowns are the position coordinates of the receiver, (x, y, z) and the local clock error, t_b. Given the actual receiver location and the actual local clock bias correction the equations below would be satisfied.

$$\left[\left(x^1 - x\right)^2 + \left(y^1 - y\right)^2 + \left(z^1 - z\right)^2\right]^{0.5} = d^1 + ct_b \tag{4.13a}$$

$$\left[\left(x^2 - x\right)^2 + \left(y^2 - y\right)^2 + \left(z^2 - z\right)^2\right]^{0.5} = d^2 + ct_b \tag{4.13b}$$

$$\left[\left(x^3 - x\right)^2 + \left(y^3 - y\right)^2 + \left(z^4 - z\right)^2\right]^{0.5} = d^3 + ct_b \tag{4.13c}$$

$$\left[\left(x^4 - x\right)^2 + \left(y^4 - y\right)^2 + \left(z^4 - z\right)^2\right]^{0.5} = d^4 + ct_b \tag{4.13d}$$

The distance calculations on the left-hand side are based on the known locations of the satellites at the time the signals were transmitted and the current estimate of the receiver location. The d^i on the right-hand side are computed from travel time of the signals multiplied by the speed of light. This pseudo-distance from each satellite to the receiver is then corrected for the local clock bias. The correct values for x, y, z, and t_b should cause all these equations to be satisfied. Here equations are shown for four satellites, the minimum required to yield the location of the receiver and the correction for the local clock bias. If more satellites are visible, more equations may be added to those below providing an even more accurate solution.

As a matter of fact, the locations of the GPS satellites are not perfectly known. Also, there are errors in the determination of time of arrival of the signals from the satellites causing the pseudo-distances to have errors. This problem, including the random errors, is an estimation problem which will be treated in a later chapter following the introduction of the Kalman Filter. For now, the problem will be treated as a deterministic problem that assumes perfect knowledge of the satellite positions and the pseudo-distances.

These equations may be rearranged to express the error between the ranges from the assumed receiver location to the respective satellites and the corrected pseudo-ranges as determined from signal time of travel.

$$E^1 = \left[\left(x^1 - x \right)^2 + \left(y^1 - y \right)^2 + \left(z^1 - z \right)^2 \right]^{0.5} - \left(d^1 + ct_b \right) \tag{4.14a}$$

$$E^2 = \left[\left(x^2 - x \right)^2 + \left(y^2 - y \right)^2 + \left(z^2 - z \right)^2 \right]^{0.5} - \left(d^2 + ct_b \right) \tag{4.14b}$$

$$E^3 = \left[\left(x^3 - x \right)^2 + \left(y^3 - y \right)^2 + \left(z^4 - z \right)^2 \right]^{0.5} - \left(d^3 + ct_b \right) \tag{4.14c}$$

$$E^4 = \left[\left(x^4 - x \right)^2 + \left(y^4 - y \right)^2 + \left(z^4 - z \right)^2 \right]^{0.5} - \left(d^4 + ct_b \right) \tag{4.14d}$$

Since the equations are nonlinear, the solution is not straightforward but rather requires an iterative process. First one makes an initial guess at the receiver location and the local clock error. Zero would be a reasonable first guess for the local clock error. Now unless one made a perfect guess of the receiver location and the local clock error, the above nonlinear equations would not be satisfied. Thus the actual position of the receiver must be determined by a series of corrections to the assumed location. One approach is to use Newton's method to force the above error vector to zero.

The equations for the errors may be written more concisely as

$$E^i = \left[\left(X - X^i \right)^T \left(X - X^i \right) \right]^{0.5} - \left(d^i + ct_b \right), \quad i = 1, ..., N \tag{4.15}$$

where the receiver location is

$$X = \begin{bmatrix} x \\ y \\ z \end{bmatrix}$$

and the *i*th satellite location is

$$X^i = \begin{bmatrix} x^i \\ y^i \\ z^i \end{bmatrix}$$

Here N is the number of visible satellites.
Defining

$$r^i = \left\{ \left(x - x^i \right)^2 + \left(y - y^i \right)^2 + \left(z - z^i \right)^2 \right\}^{0.5}$$

the above equations simplify to

$$E^i = r^i - \left(d^i + ct_b\right), \quad i = 1, \ldots, N \tag{4.16}$$

One can now expand the error equation using the Taylor series through the linear terms to obtain

$$E + \Delta E = \begin{bmatrix} r^1 - \left(d^1 + ct_b\right) \\ r^2 - \left(d^2 + ct_b\right) \\ \vdots \\ r^N - \left(d^N + ct_b\right) \end{bmatrix} + [\partial E/\partial X \; \partial E/\partial ct_b] \begin{bmatrix} \Delta x \\ \Delta y \\ \Delta z \\ \Delta ct_b \end{bmatrix} \tag{4.17}$$

Differentiating the equation for the error yields

$$\partial E^i/\partial X = \left[\left(X - X^i\right)^T \left(X - X^i\right)\right]^{-0.5} \left(X - X^i\right)^T = 1/r^i \left(X - X^i\right)^T$$

and

$$\partial E^i/\partial ct_b = -1$$

Note that this last derivative is with respect to ct_b, which has dimensions of distance rather than with respect to t_b, which has dimension of time. The reason for this is that the other derivatives of the error were taken with respect to distance. Doing this removes one potential source of numerical errors, large differences in scale. In expanded form, this becomes

$$E + \Delta E = \begin{bmatrix} r^1 - \left(d^1 + ct_b\right) \\ r^2 - \left(d^2 + ct_b\right) \\ \vdots \\ r^N - \left(d^N + ct_b\right) \end{bmatrix}$$

$$+ \begin{bmatrix} \frac{1}{r^1}\left(x - x^1\right) & \frac{1}{r^1}\left(y - y^1\right) & \frac{1}{r^1}\left(z - z^1\right) & -1 \\ \frac{1}{r^2}\left(x - x^2\right) & \frac{1}{r^2}\left(y - y^2\right) & \frac{1}{r^2}\left(z - z^2\right) & -1 \\ \vdots & \vdots & \vdots & \vdots \\ \frac{1}{r^N}\left(x - x^N\right) & \frac{1}{r^N}\left(y - y^N\right) & \frac{1}{r^N}\left(z - z^N\right) & -1 \end{bmatrix} \begin{bmatrix} \Delta x \\ \Delta y \\ \Delta z \\ \Delta ct_b \end{bmatrix}$$

$$\tag{4.18}$$

One seeks the appropriate values for ΔE to force this error to zero. Setting $E + \Delta E = 0$ and solving for the changes in estimated receiver location and receiver clock bias yields

$$
\begin{bmatrix} \Delta x \\ \Delta y \\ \Delta z \\ \Delta ct_b \end{bmatrix} = -\left[\partial E/\partial X \quad \partial E/\partial ct_b \right]^{-1} \begin{bmatrix} r^1 - \left(d^1 + ct_b\right) \\ r^2 - \left(d^2 + ct_b\right) \\ \vdots \\ r^N - \left(d^N + ct_b\right) \end{bmatrix}
\tag{4.19}
$$

or

$$
\begin{bmatrix} \Delta x \\ \Delta y \\ \Delta z \\ \Delta ct_b \end{bmatrix} = -
\begin{bmatrix}
\frac{1}{r^1}\left(x - x^1\right) & \frac{1}{r^1}\left(y - y^1\right) & \frac{1}{r^1}\left(z - z^1\right) & -1 \\
\frac{1}{r^2}\left(x - x^2\right) & \frac{1}{r^2}\left(y - y^2\right) & \frac{1}{r^2}\left(z - z^2\right) & -1 \\
\vdots & \vdots & \vdots & \vdots \\
\frac{1}{r^N}\left(x - x^N\right) & \frac{1}{r^N}\left(y - y^N\right) & \frac{1}{r^N}\left(z - z^N\right) & -1
\end{bmatrix}^{-1}
$$

$$
\begin{bmatrix} r^1 - \left(d^1 + ct_b\right) \\ r^2 - \left(d^2 + ct_b\right) \\ \vdots \\ r^N - \left(d^N + ct_b\right) \end{bmatrix}
\tag{4.20}
$$

In practice, a scale factor less than one is often introduced to aid in the convergence of the solution. For cases where the number of visible satellites is greater than four, there will be more equations than unknowns. To proceed, one may utilize the least-squares solution in which case the matrix inverse becomes a generalized inverse.

$$
\begin{bmatrix} \Delta x \\ \Delta y \\ \Delta z \\ \Delta ct_b \end{bmatrix} = \left\{ \left[\partial E/\partial X \quad \partial E/\partial ct_b \right]^T \left[\partial E/\partial X \quad \partial E/\partial ct_b \right] \right\}^{-1} \left[\partial E/\partial X \quad \partial E/\partial ct_b \right]^T \begin{bmatrix} r^1 - \left(d^1 + ct_b\right) \\ r^2 - \left(d^2 + ct_b\right) \\ \vdots \\ r^N - \left(d^N + ct_b\right) \end{bmatrix}
\tag{4.21}
$$

Fortunately, these equations are all consistent, which means that there exists an exact solution (in the absence of noise) even though the number of equations exceeds the number of unknowns. After solving equation (4.21), one updates the receiver location and the local clock bias correction according to

$$
\begin{bmatrix} x \\ y \\ z \\ ct_b \end{bmatrix} = \begin{bmatrix} x \\ y \\ z \\ ct_b \end{bmatrix} + \begin{bmatrix} \Delta x \\ \Delta y \\ \Delta z \\ \Delta ct_b \end{bmatrix} \tag{4.22}
$$

This correction process is repeated until it reaches steady state, i.e., the correction approaches zero. At this time the distances computed from the equations involving receiver location should agree with the pseudo-distances based on the measurements of signal travel time with local clock bias correction.

Example 6 *At a given point, signals are received from four different satellites. At the time the signals left the satellites, their positions expressed in ECEF coordinates were as follows:*

$$x1 = 7,766,188.44;$$
$$y1 = -21,960,535.34;$$
$$z1 = 12,522,838.56;$$
$$x2 = -25,922,679.66;$$
$$y2 = -6,629,461.28;$$
$$z2 = 31,864.37;$$
$$x3 = -5,743,774.02;$$
$$y3 = -25,828,319.92;$$
$$z3 = 1,692,757.72;$$
$$x4 = -2,786,005.69;$$
$$y4 = -15,900,725.8;$$
$$z4 = 21,302,003.49;$$

Based on the times the signals were received and the information regarding when they were transmitted from the respective satellites, their pseudo-distances from the receiver are computed to be

$$d1 = 1,022,228,206.42$$
$$d2 = 1,024,096,139.11$$

$$d3 = 1,021,729,070.63$$
$$d4 = 1,021,259,581.09$$

What is the receiver location?

Solution 6

A program was developed for solving this problem on the computer. The initial guess for the receiver location is the origin of the ECEF coordinate system and the initial guess for local clock error is zero.

```
N = 10;
C = 3 * 10^8;

d1 = 1022228206.42; x1 = 7766188.44;
y1 = -21960535.34; z1 = 12522838.56;
X1 = [x1 y1 z1]';

d2 = 1024096139.11; x2 = -25922679.66;
y2 = -6629461.28; z2 = 31864.37;
X2 = [x2 y2 z2]';

d3 = 1021729070.63; x3 = -5743774.02;
y3 = -25828319.92; z3 = 1692757.72;
X3 = [x3 y3 z3]';

d4 = 1021259581.09; x4 = -2786005.69;
y4 = -15900725.8; z4 = 21302003.49;
X4 = [x4 y4 z4]';

x = 0; y = 0; z = 0;
X = [x y z]';

ctb = 0; delX = 0; delctb = 0;

for j = 1:N

    X = X + delX;
    ctb = ctb + delctb;
    r1 = ((X - X1)' * (X - X1))^.5;
    r2 = ((X - X2)' * (X - X2))^.5;
    r3 = ((X - X3)' * (X - X3))^.5;
    r4 = ((X - X4)' * (X - X4))^.5;
```

```
    error = -[r1 - (d1 + ctb); r2 - (d2 + ctb);
            r3 - (d3 + ctb); r4 - (d4 + ctb)];
%These are the errors between distances computed
based on respective satellite positions and
%current estimated receiver position versus
distances computed based on the times the signals
were
%received from the respective satellites

    L(j) = (r1 - (d1 + ctb))^2 + (r2 - (d2 + ctb))^2
            + (r3 - (d3 + ctb))^2 + (r4 - (d4 + ctb))^2;
%positive definite measure of error

    grad1 = [-(1 / r1) * (X1 - X)' -1];
    grad2 = [-(1 / r2) * (X2 - X)' -1];
    grad3 = [-(1 / r3) * (X3 - X)' -1];
    grad4 = [-(1 / r4) * (X4 - X)' -1];
    Grad = [grad1; grad2; grad3; grad4];
    Gradinv = (inv(Grad' * Grad)) * Grad';
% A generalized inverse is used in order to be
able to handle cases where more than 4 satellites
are visible.

    delX = K * [1 0 0 0; 0 1 0 0; 0 0 1 0] * Gradinv *
error;
    delctb = K * [0 0 0 1] * Gradinv * error;
%This is the correction desired at each step,
K is often set to be less than one, maybe 0.5 or
so,  to
%improve convergence and reduce the probability
of overshoot in the solution process.

end
figure
plot(L)
```

The output of the program was the receiver location which is

$x = -2,430,745$

$y = -4,702,345$

$z = 3,546,569$

and the local clock error

$tb = -3.3342156$

The extremely rapid convergence of x, y, z, and total error can be seen from Figure 4.3a and b as well as from Table 4.1. It is seen that after five iterations, the solution has been reached within half a meter in each coordinate. The value used for K was 1.0.

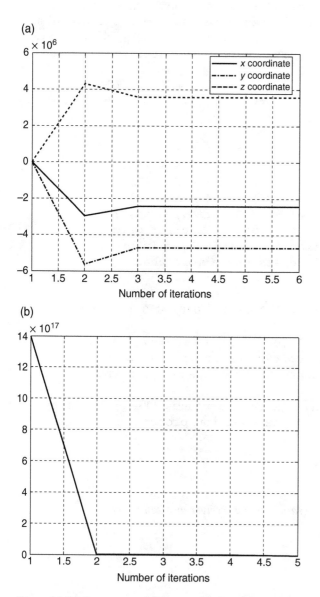

Figure 4.3 (a) Convergence of the coordinates of the receiver position. (b) Convergence of the norm of GPS error. Newton method was used.

Table 4.1 Convergence of coordinates as a function of iteration number. Newton method was used.

Iteration	1	2	3	4	5	50
Performance	3.9658	1.3970	2.1109	3.1469	7.3558	1.3878
Index	e+018	e+012	e+009	e+003	e−009	e−015
x	0	−2,977,571	2,451,728	−2,430,772	−2,430,745	−2,430,745
y	0	−5,635,278	−4,730,878	−4,702,376	−4,702,345	−4,702,345
z	0	4,304,234	3,573,997	3,546,603	3,546,569	3,546,569
ctb	0	−3.338750	−3.3343802	−3.334215	−3.334215	−3.3342156

If the receiver is mounted on a moving vehicle, its position may be tracked by successive GPS measurements. Once its first location has been determined, its next location can be determined quite rapidly since the previous location serves as a good initial guess for the iterative process. Probably only one iteration would be required.

4.7.2 Computing Receiver Location Using GPS via Minimization of a Performance Index

An alternative approach to determining the coordinates of the receiver is to formulate a positive definite performance index based on the sum of squares of the distance errors described in equation (4.16). The performance measure used to represent total error in geolocation will be taken as

$$L = \sum_{i=1}^{N} \left\{ \left[\left(x - x^i \right)^2 + \left(y - y^i \right)^2 + \left(z - z^i \right)^2 \right]^{0.5} - \left[d^i + ct_b \right] \right\}^2 \qquad (4.23)$$

where, as was the case before, the coordinates of the ith satellite at time of transmission are given by the coordinates $[x^i, y^i, z^i]^T$ and the coordinates of the receiver are initially estimated to be $[x, y, z]^T$. N is the number of visible satellites. Utilizing $X^i = [x^i, y^i, z^i]^T$ and $X = [x, y, z]^T$, the above can be written in shorthand as

$$L = \sum_{i=1}^{N} \left\{ \left[\left(X - X^i \right)^T \left(X - X^i \right) \right]^{0.5} - \left[d^i + ct_b \right] \right\}^2$$

or

$$L = \sum_{i=1}^{N} \left\{ r^i - \left[d^i + ct_b \right] \right\}^2 \qquad (4.24)$$

It is seen that the individual error terms are squared before being added together. Otherwise negative and positive errors could cancel each other, leading to a small value of L even though there were significant errors. Note that this is the same expression used in the Matlab program of the previous example to monitor the progress of the Newton iterative method. Here this performance index will be used to actually define the iterative procedure. The goal is to determine the coordinates of the receiver which cause this performance index to be minimized. Starting with an initial guess, one successively perturbs the coordinates of the receiver and the local clock error to move the solution toward this minimum. As part of the numerical search procedure, the gradient of the performance index must be determined in order to know in what direction to perturb the coordinates of the receiver. Differentiating the above expression of the performance index, one obtains

$$\partial L / \partial X = \sum_{i=1}^{N} 2 \left\{ \left[(X - X^i)^T (X - X^i) \right]^{0.5} - (d^i + ct_b) \right\} \times \\ \frac{1}{2} \left[(X - X^i)^T (X - X^i) \right]^{-0.5} 2 (X - X^i)^T$$

Using the definition for r^i the above derivative simplifies to

$$\partial L / \partial X = 2 \sum_{i=1}^{N} \left\{ r^i - (d^i + ct_b) \right\} (X - X^i)^T / r^i \tag{4.25}$$

This vector points in the direction of increasing L. The goal here is to minimize L, so one is interested in the direction of decrease, i.e.,

$$- \partial L / \partial X = 2 \sum_{i=1}^{N} \left\{ r^i - (d^i + ct_b) \right\} (X^i - X)^T / r^i$$

Note that the portion $(X^i - X)/r^i$ is a unit vector pointing from the current estimate of the receiver location to the location of the ith satellite. The weighting for this ith component of $-(\partial L / \partial X)$ is the error in distances as calculated from the current estimate of the receiver position and the ith satellite position and the time-corrected pseudo-distance from the receiver to the ith satellite. If the distance as calculated from the current estimate of the receiver position and the ith satellite position is greater than the time-corrected pseudo-distance from the receiver to the ith satellite, then this component of $-(\partial L / \partial X)$ would indicate that one should perturb the estimated receiver location toward the ith satellite. For the opposite situation, the perturbation would be to move the estimated receiver location away from the ith satellite. There will be N such vectors added together, each weighed in proportion to the respective errors and each moving toward or away from the respective satellites. The perturbations will be confined to the space spanned by the vectors from the receiver to the visible satellites. A set of satellites having a low value for GDOP, i.e., having a good geometric distribution, will result in a more robust iterative process. If the visible satellites were all in the same plane, corrections in receiver position using this iterative process would be limited to this plane.

One also needs to adjust the local clock bias. The result of this differentiation becomes

$$\partial L/\partial ct_b = \sum_{i=1}^{N} 2\{r^i - (d^i + ct_b)\}(-1) \tag{4.26}$$

and again the negative value is of interest since the goal is to minimize L,

$$-\partial L/\partial ct_b = \sum_{i=1}^{N} 2\{r^i - (d^i + ct_b)\}$$

This direction indicated by the objective of decreasing L is seen to be that of an increase in ctb if the computed range is greater than the corrected pseudo-range. One uses these derivative functions as a means of determining the perturbation of the four-dimensional vector $[x, y, z, ct_b]^T$. One must also know what step size to use for the perturbation. Oftentimes, the function to be minimized is very complex and a step size too large may overshoot the solution or a step size too small may take very long to converge. Also, one step size may not apply everywhere. A means of determining not only the direction but also the step size for the iterative process is to utilize a more complete Taylor series including the second derivative of the function to be minimized, i.e.,

$$f(w + \Delta w) \approx f(w) + \partial f/\partial w \Delta w + \frac{1}{2}\Delta w^T \left[\partial(\partial f/\partial w)^T/\partial w\right]\Delta w$$

Then minimizing

$$f(w + \Delta w)$$

with respect to Δw yields

$$\Delta w = -\left[\frac{\partial}{\partial w}\left\{(\partial f/\partial w)^T\right\}\right]^{-1}(\partial f/\partial w)^T \tag{4.27}$$

This may be interpreted as using the Newton method to determine the point where the first derivative is zero. Performing these operations on the performance index of interest, one obtains as the matrix of second derivatives

$$\frac{\partial}{\partial X}\left\{(\partial L/\partial X)^T\right\} = 2\sum_{i=1}^{N}\left\{\frac{r^i - (d^i + ct_b)}{r^i}I + \frac{(d^i + ct_b)}{(r^i)^3}(X - X^i)(X - X^i)^T\right\}$$

$$\frac{\partial}{\partial X}(\partial L/\partial ct_b) = -2\sum_{i=1}^{N}\frac{1}{r^i}(X - X^i)^T$$

$$\frac{\partial}{\partial ct_b}(\partial L/\partial X)^T = -2\sum_{i=1}^{N}\frac{1}{r^i}\left(X-X^i\right)$$

and

$$\partial^2 L/\partial ct_b^2 = 2N$$

The changes in X and ctb then become

$$\begin{bmatrix} \Delta X \\ \Delta ctb \end{bmatrix} = -\begin{bmatrix} \dfrac{\partial}{\partial X}(\partial L/\partial X)^T & \dfrac{\partial}{\partial ct_b}(\partial L/\partial X)^T \\[2ex] \dfrac{\partial}{\partial X}(\partial L/\partial ct_b) & \dfrac{\partial}{\partial ct_b}(\partial L/\partial ct_b) \end{bmatrix}^{-1}\begin{bmatrix} (\partial L/\partial X)^T \\[2ex] \partial L/\partial ct_b \end{bmatrix} \qquad (4.28)$$

Example 7 *Solve the previous geolocation problem, this time via minimizing a performance index.*

Solution 7

Below is shown the computer code for using the above described minimization procedure to determine the receiver coordinates. The data on satellite positions and distances are the same as before and are omitted here to save space.

```
                        N = 50;
delX = 0; delctb = 0;   ctb = 0;
x = 0;  y = 0;  z = 4000000;
X = [x y z]';

for j = 1:N

    X = X + delX;
    ctb = ctb + delctb;
    r1 = ((X - X1)' * (X - X1))^.5;
    r2 = ((X - X2)' * (X - X2))^.5;
    r3 = ((X - X3)' * (X - X3))^.5;
    r4 = ((X - X4)' * (X - X4))^.5;

    L(j) = (r1 -(d1 + ctb))^2 + (r2 - (d2 +
           ctb))^2 + (r3 -(d3 + ctb))^2 + (r4
           - (d4 + ctb))^2;
```

```
delLx = (2 / r1) * (r1 - (d1 + ctb)) * (X
        - X1) + (2 / r2) * (r2 - (d2 +
        ctb)) * (X - X2)
        + (2 / r3) * (r3 - (d3 + ctb)) * (X - X3)
        + (2 / r4) * (r4 - (d4 + ctb)) * (X - X4);
delLctb = -2 * ((r1 - (d1 + ctb)) + (r2 - (d2
          + ctb)) + (r3 - (d3 + ctb)) + (r4
          - (d4 + ctb)));
delLxdelLx = 2 * ((r1 - (d1 + ctb)) / r1
             + (r2 - (d2 + ctb)) / r2 + (r3
             - (d3 + ctb)) / r3
             + (r4 - (d4 + ctb)) / r4) * eye(3);
delLxdelLx = delLxdelLx + 2 * (((d1 + ctb) / r1^3)
             * (X - X1) * (X - X1)'
             + ((d2 + ctb) / r2^3) * (X - X2)
             * (X - X2)'
             + ((d3 + ctb) / r3^3) * (X - X3)
             * (X - X3)'
             + ((d4 + ctb) / r4^3) * (X - X4)
             * (X - X4)');
delLctbdelLx = -(2 / r1) * (X - X1) - (2 / r2) *
               (X - X2) - (2 / r3) * (X - X3) - (2 / r4)
               * (X - X4);
delLxdelLctb = -(2 / r1) * (X - X1)' - (2 / r2)
               * (X - X2)' - (2 / r3) * (X - X3)'
               - (2 / r4) * (X - X4)';
delLctbdelLctb = 2 * 4;

D2 = [delLxdelLx delLctbdelLx; delLxdelLctb
      delLctbdelLctb];
delX = -[1 0 0 0; 0 1 0 0; 0 0 1 0] *
       inv(D2) * [delLx; delLctb];
delctb = -[0 0 0 1] * inv(D2) * [delLx; delLctb];
xx(j) = [1 0 0] * X;
yy(j) = [0 1 0] * X;
zz(j) = [0 0 1] * X;
ctbb(j) = ctb;
end

figure
plot(L)
axis([1 10 0 5*10^18])
```

```
figure
plot(xx); hold on; plot(yy); hold on; plot(zz)
axis([1 10 -1.5*10^7 1*10^7])
```

From the code it is seen that the procedure was started with initial guess of
$x = 0$, $y = 0$, *and* $z = 4,000,000$. *It had been found that the iterative process converged to a wrong solution unless one placed the initial value of z in the correct hemisphere. The rapid convergence of x, y, z, and total error can be seen from Figure 4.4a and b as well as from Table 4.2. It is seen that after 10 iterations the solution has been reached within half a meter in each coordinate. This was not as fast as the convergence of the first method, but nevertheless quite fast.*

Either of the two methods presented would be sufficient for initialization, and either method would probably converge in one step while tracking a moving vehicle using the previous location as the first guess for the new location. The first method is considerably simpler in terms of analytical and computational complexity as well as having faster convergence.

(a)

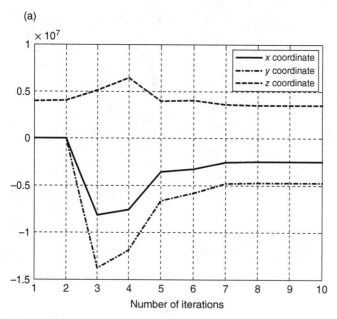

Figure 4.4 (a) Convergence of the coordinates of the receiver position. (b) Convergence of the norm of GPS error. Performance Measure minimization was used.

(b)

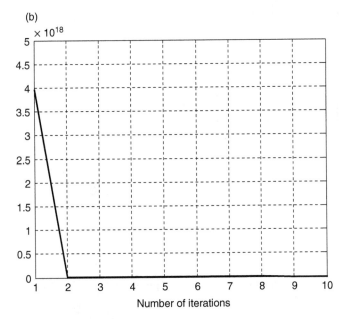

Figure 4.4 (Continued)

Table 4.2 Convergence of coordinates as a function of iteration number. Performance Measure minimization was used.

Iteration	1	3	6	8	10	50
Performance Index	3.9745 e+018	7.271354 e+013	1.5179206 e+011	5.2192730 e+006	7.0935216 e−011	1.3877787 e−015
x	0	−8,150,229	−3,247,551	−2,435,857	−2,430,745	−2,430,745.1
y	0	−1,381,163	−5,823,740	−4,708,767	−4,702,345	−4,702,345.1
z	4e+6	5,136,466	4,015,697	3,550,007	3,546,568	3,546,568.7
ctb	0	−3.338750	−3.3372523	−3.334235	−3.334215	−3.33421563

4.8 Array of GPS Antennas

An array of four GPS antennas, such as shown in Figure 4.5, may be used to compute vehicle attitude. These are attached to a frame in the shape of a cross with antennas labeled 1 (front), 2 (left side), 3 (rear), and 4 (right side). The x axis runs across to the right side of the frame. The y axis runs from the rear to the front of the frame. The z axis completes the right-handed set.

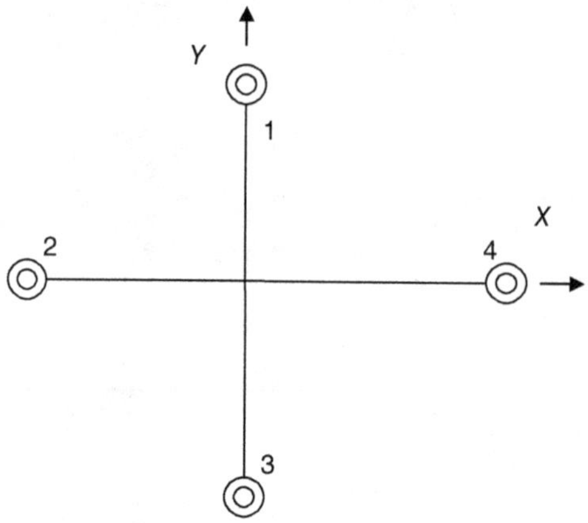

Figure 4.5 Convergence of coordinates as a function of iteration number array of four GPS antennas.

Before any rotation of the vehicle, i.e., zero yaw, pitch, and roll, the coordinates are

$$x_1 = x_{veh}, \quad y_1 = L/2 + y_{veh}, \quad z_1 = z_{veh} \tag{4.29a}$$

$$x_2 = -W/2 + x_{veh}, \quad y_2 = y_{veh}, \quad z_2 = z_{veh} \tag{4.29b}$$

$$x_3 = x_{veh}, \quad y_3 = -L/2 + y_{veh}, \quad z_3 = z_{veh} \tag{4.29c}$$

and

$$x_4 = W/2 + x_{veh}, \quad y_4 = y_{veh}, \quad z_4 = z_{veh} \tag{4.29d}$$

Now after rotation of the vehicle through a yaw angle of ψ, a pitch angle of θ, and a roll angle of ϕ in that order, the coordinates become

$$x_1 = -L/2(\sin\psi\,\cos\theta) + x_{veh} \tag{4.30a}$$

$$y_1 = L/2(\cos\psi\,\cos\theta) + y_{veh} \tag{4.30b}$$

$$z_1 = L/2(\sin\theta) + z_{veh} \tag{4.30c}$$

$$x_2 = -W/2(\cos\psi\,\cos\phi - \sin\psi\,\sin\theta\,\sin\phi) + x_{veh} \tag{4.31a}$$

$$y_2 = -W/2(\sin\psi\,\cos\phi + \cos\psi\,\sin\theta\,\sin\phi) + y_{veh} \tag{4.31b}$$

$$z_2 = W/2(\cos\theta\sin\phi) + z_{veh} \tag{4.31c}$$

$$x_3 = L/2(\sin\psi\cos\theta) + x_{veh} \tag{4.32a}$$

$$y_3 = -L/2(\cos\psi\cos\theta) + y_{veh} \tag{4.32b}$$

$$z_3 = -L/2(\sin\theta) + z_{veh} \tag{4.32c}$$

and

$$x_4 = W/2(\cos\psi\cos\phi - \sin\psi\sin\theta\sin\phi) + x_{veh} \tag{4.33a}$$

$$y_4 = W/2(\sin\psi\cos\phi + \cos\psi\sin\theta\sin\phi) + y_{veh} \tag{4.33b}$$

$$z_4 = -W/2(\cos\theta\sin\phi) + z_{veh} \tag{4.33c}$$

By manipulating these equations, it is possible to isolate the attitude angles in terms of the measured variables. For pitch we have

$$\theta = \tan^{-1}\left[\frac{(z_1 - z_3)}{\sqrt{(x_1 - x_3)^2 + (y_1 - y_3)^2}}\right] \tag{4.34}$$

Yaw is given by

$$\psi = \tan^{-1}\left[\frac{-(x_1 - x_3)}{(y_1 - y_3)}\right] \tag{4.35}$$

For roll

$$\phi = \sin^{-1}\left[\frac{(z_2 - z_4)}{W\cos\theta}\right] = \sin^{-1}\left[\frac{(z_2 - z_4)/W}{\sqrt{(x_1 - x_3)^2 + (y_1 - y_3)^2/L}}\right] \tag{4.36}$$

Notice that in all these expressions the coordinates of the vehicle location subtract off and do not affect the expressions for vehicle attitude. The major errors in the measurements at antenna locations will be common to all four antennas. Thus, even though absolute positioning of the antennas may not be possible, the computed attitude may be extremely accurate. As an example, one GPS array unit advertises accuracy in position of only 40 cm but accuracy in attitude of less than six-tenths of a degree.

4.9 Gimbaled Inertial Navigation Systems

The two distinct implementation approaches for INS will be discussed. These are gimbaled and strap-down gyro systems. For the gimbaled system there is an actuated platform on which are mounted three orthogonal gyros. A gimbaled platform is shown in Figure 4.6.

Here the gyros on the platform in conjunction with the gimbal motors maintain the platform at a fixed attitude in an inertial frame even though the mounting frame, which is attached to the vehicle of interest, may rotate. Thus the platform is called a stable platform. In studying the behavior of the gyros, we use one of the equations that governs the behavior of a rotating body, i.e.,

$$\tau = \Omega \times L \tag{4.37}$$

Here L is the angular momentum of the gyro and it points along the spin axis. The vector Ω is the angular precession velocity of this momentum vector. The associated torque is represented by the vector τ. As the attitude of the vehicle changes, a small angular velocity, Ω of the platform and thus an equal angular rotation of the gyro angular momentum vector may occur as a result of slight friction in the imperfect gimbal bearings. The torque, τ resulting from this rotation according to equation (4.37) is sensed at the gyro bearing supports. Through feedback control the gimbal motors react to negate this torque maintaining the platform at the fixed attitude. The torques will be sensed by the gyros in platform coordinates; however, the gimbal motors will be aligned according to the attitude of the vehicle. Thus a coordinate transformation is necessary

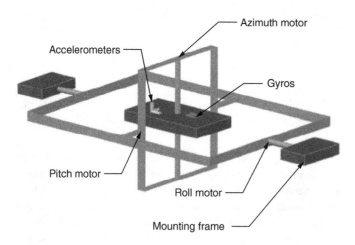

Figure 4.6 Schematic diagram of a gimbaled platform.

to determine the required gimbal rotations. Through this arrangement the platform is maintained at a fixed attitude even though the vehicle may rotate. Shaft encoders or similar sensors are used to measure the gimbal angles and thus provide the current attitude of the vehicle with respect to the stable platform. From the figure it can be seen that there is a gimbal and a gimbal motor each for yaw (or azimuth), pitch, and roll.

In addition to the gyros, three accelerometers are also mounted on the stable platform. They read the three components of acceleration in inertial coordinates. By integrating these signals twice with respect to time, one obtains change in position in all three coordinates. Combining this information with the original position gives one the current position of the vehicle in inertial space. If there is any offset or bias in the accelerometers, the process of double integration creates an error that grows as the square of time, eventually becoming unacceptable. Some auxiliary means of determining position is required for periodic re-calibration of the inertial unit. The time between re-calibration is dictated by the accuracy requirements and the extent of the accelerometer bias. Biases may range from a few hundredths of a milligram to a few milligrams.

In addition to these position errors caused by accelerometer bias, there are attitude errors caused by drift, i.e., the system thinks it is rotating when it is not. The drift rates may be as high as a few degrees per hour and as low as a few milli-degrees per hour.

Errors in attitude do not grow as fast as errors in position since attitude estimates do not suffer from the double integration of biases as the position estimates do.

Example 8 *An accelerometer yields acc(indicated) = acc + n + b m/s², where acc represents the true acceleration, n represents random errors of zero mean, and b represents a bias. Let b = 10^{-5}m/s² or 1.02×10^{-3}mg. Compute position from this signal. If the maximum position error one can tolerate is 1 m, at what time intervals must the computations be corrected with an independent, correct position measurement?*

Solution 8

The integration of the accelerometer output yields as vel(t), the indicated velocity,

$$vel(t) = \int_0^t (acc + n + b)dt + v(0)$$

$$vel(t) = \int_0^t accdt + v(0) + \int_0^t bdt + \int_0^t ndt$$

$$vel(t) = v(t) + bt + 0$$

where $v(t)$ represents the true velocity and bt represents the error caused by the bias. The error caused by the random noise term integrates to zero since it has zero mean.

$$Position(t) = \int_0^t (vel(t))dt + P(0)$$

or

$$Position(t) = \int_0^t v(t)dt + \int_0^t btdt + P(0)$$

$$Position(t) = P(t) + b\frac{t^2}{2}$$

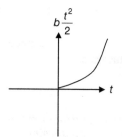

Now setting the error term to its maximum allowable value

$$b\frac{t^2}{2} = 1\,m$$

yields

$$t^2 = \frac{2}{b}$$

or

$$t = \sqrt{\frac{2}{b}}$$

$$t = \sqrt{2(10^5)} = 447.213\ seconds = 7.45\ minutes$$

Next the question of obtaining vehicle attitude and position from discrete-time outputs of the gimbaled gyroscope with stable platform is addressed. When using the gimbaled gyroscope, the attitude equations are simply:

for yaw

$$\psi(t_{k+1}) = \psi_{measured}(t_{k+1})$$

for pitch

$$\theta(t_{k+1}) = \theta_{measured}(t_{k+1})$$

and for roll

$$\phi(t_{k+1}) = \phi_{measured}(t_{k+1})$$

In other words, the measurements of the gimbal angles are the same as the attitude measurements of the vehicle with respect to the stable platform. For position the equations are:

for x

$$\dot{x}(t_{k+1}) = \dot{x}(t_k) + a_x(t_k)_{measured}(t_{k+1} - t_k) \tag{4.38a}$$

$$x(t_{k+1}) = x(t_k) + \dot{x}(t_{k+1})(t_{k+1} - t_k) + a_x(t_k)_{measured}\frac{(t_{k+1} - t_k)^2}{2} \tag{4.38b}$$

for y

$$\dot{y}(t_{k+1}) = \dot{y}(t_k) + a_y(t_k)_{measured}(t_{k+1} - t_k) \tag{4.38c}$$

$$y(t_{k+1}) = y(t_k) + \dot{y}(t_k)(t_{k+1} - t_k) + a_y(t_k)_{measured}\frac{(t_{k+1} - t_k)^2}{2} \tag{4.38d}$$

and for z

$$\dot{z}(t_{k+1}) = \dot{z}(t_k) + a_z(t_k)_{measured}(t_{k+1} - t_k) \tag{4.38e}$$

$$z(t_{k+1}) = z(t_k) + \dot{z}(t_k)(t_{k+1} - t_k) + a_z(t_k)_{measured}\frac{(t_{k+1} - t_k)^2}{2} \tag{4.38f}$$

Attitude and position are expressed in inertial space, i.e., a nonrotating, non-moving coordinate system. Here the assumption of constant acceleration over the sampling interval has been assumed, i.e.,

$$a(t) = a(t_k)_{measured} \quad \text{for} \quad t_k \leq t < t_{k+1}$$

A combination of $a(t_k)_{measured}$ and $a(t_{k+1})_{measured}$ could be used for better accuracy if one is willing to wait for the measurement at time t_{k+1}.

4.10 Strap-Down Inertial Navigation Systems

In addition to gimbaled INS, there are also strap-down systems. Here the spin axes for the gyros are rigidly attached to the vehicle, which means that they change in attitude as the vehicle changes attitude. Since the sensors experience the full dynamic motion of the vehicle, higher bandwidth rate gyros with higher dynamic range are required. The equation relating precession rate to torque $\tau = \Omega \times L$ can be broken down into components

$$
\begin{bmatrix} \tau_x \\ \tau_y \\ \tau_z \end{bmatrix} = \begin{bmatrix} 0 & L_z & -L_y \\ -L_z & 0 & L_x \\ L_y & -L_x & 0 \end{bmatrix} \begin{bmatrix} \Omega_x \\ \Omega_y \\ \Omega_z \end{bmatrix}
\tag{4.39}
$$

There will be an equation of this type for each of the three orthogonal gyros. For the gyro aligned with the x axis of the platform, only L_x is nonzero, and τ_x is zero leading to

$$
\begin{bmatrix} \tau_y \\ \tau_z \end{bmatrix}_{gyro-x} = \begin{bmatrix} 0 & L_x \\ -L_x & 0 \end{bmatrix} \begin{bmatrix} \Omega_y \\ \Omega_z \end{bmatrix}
\tag{4.40a}
$$

For the gyro aligned with the y axis of the platform, only L_y is nonzero, and τ_y is zero leading to

$$
\begin{bmatrix} \tau_x \\ \tau_z \end{bmatrix}_{gyro-y} = \begin{bmatrix} 0 & -L_y \\ L_y & 0 \end{bmatrix} \begin{bmatrix} \Omega_x \\ \Omega_z \end{bmatrix}
\tag{4.40b}
$$

For the gyro aligned with the z axis of the platform, only L_z is nonzero, and τ_z is zero leading to

$$
\begin{bmatrix} \tau_x \\ \tau_y \end{bmatrix}_{gyro-z} = \begin{bmatrix} 0 & L_z \\ -L_z & 0 \end{bmatrix} \begin{bmatrix} \Omega_x \\ \Omega_y \end{bmatrix}
\tag{4.40c}
$$

These overdetermined equations (six equations and three unknowns) involve the angular momentum of each gyro (known quantities), and the six torques, which are measured at the individual gyro bearing supports.

$$
\begin{bmatrix} \tau_{y-gyro\ x} \\ \tau_{z-gyro\ x} \\ \tau_{x-gyro\ y} \\ \tau_{z-gyro\ y} \\ \tau_{x-gyro\ z} \\ \tau_{y-gyro\ z} \end{bmatrix} = \begin{bmatrix} 0 & 0 & L_x \\ 0 & -L_x & 0 \\ 0 & 0 & -L_y \\ L_y & 0 & 0 \\ 0 & L_z & 0 \\ -L_z & 0 & 0 \end{bmatrix} \begin{bmatrix} \Omega_x \\ \Omega_y \\ \Omega_z \end{bmatrix}
\tag{4.41}
$$

The solutions Ω_x, Ω_y, and Ω_z are the instantaneous angular body rates: pitch rate, roll rate, and yaw rate may be obtained via the least-squares method as

$$\Omega_x = \frac{1}{L_y^2 + L_z^2}\left(L_y \tau_{z,gyro-y} - L_z \tau_{y,gyro-z}\right) \tag{4.42a}$$

$$\Omega_y = \frac{1}{L_x^2 + L_z^2}\left(-L_x \tau_{z,gyro-x} + L_z \tau_{x,gyro-z}\right) \tag{4.42b}$$

and

$$\Omega_z = \frac{1}{L_x^2 + L_y^2}\left(L_x \tau_{y,gyro-x} - L_y \tau_{x,gyro-y}\right) \tag{4.42c}$$

Although this is a least squares solution, the equations are consistent and there is no error in the solution for the body rates. These rates are related to the attitude rates according to the following equations:

$$\Omega_x = \dot{\theta}\cos\phi + \dot{\psi}\sin\phi\cos\theta \tag{4.43a}$$

$$\Omega_y = \dot{\phi} - \dot{\psi}\sin\theta \tag{4.43b}$$

$$\Omega_z = \dot{\psi}\cos\phi\cos\theta - \dot{\theta}\sin\phi \tag{4.43c}$$

In matrix form these become

$$\begin{bmatrix} \Omega_x \\ \Omega_y \\ \Omega_z \end{bmatrix} = \begin{bmatrix} \cos\phi & 0 & \sin\phi\cos\theta \\ 0 & 1 & -\sin\theta \\ -\sin\phi & 0 & \cos\phi\cos\theta \end{bmatrix} \begin{bmatrix} \dot{\theta} \\ \dot{\phi} \\ \dot{\psi} \end{bmatrix} \tag{4.44}$$

which may be inverted to yield the angular rates for yaw, pitch, and roll.

$$\begin{bmatrix} \dot{\theta} \\ \dot{\phi} \\ \dot{\psi} \end{bmatrix} = \begin{bmatrix} \cos\phi & 0 & -\sin\phi \\ \sin\phi\tan\theta & 1 & \cos\phi\tan\theta \\ \sin\phi/\cos\theta & 0 & \cos\phi/\cos\theta \end{bmatrix} \begin{bmatrix} \Omega_x \\ \Omega_y \\ \Omega_z \end{bmatrix} \tag{4.45}$$

Once these angular rates for attitude have been determined, the attitude angles themselves can be updated via numerical integration. Using the Euler integration method, i.e., derivative approximated by forward difference, yields

$$\psi(t + \Delta t) = \psi(t) + \dot{\psi}(t)\Delta t$$

$$\theta(t + \Delta t) = \theta(t) + \dot{\theta}(t)\Delta t$$

and

$$\phi(t + \Delta t) = \phi(t) + \dot{\phi}(t)\Delta t$$

Another approach to the problem of obtaining vehicle orientation from body angular rates involves the use of quaternions. The quaternion vector is defined by

$$
q = \begin{bmatrix} q_0 \\ q_1 \\ q_2 \\ q_3 \end{bmatrix}
\tag{4.46}
$$

where the components of the quaternion are related to the vehicle attitude by

$$
q_0 = \cos{(\psi/2)} \cos{(\theta/2)} \cos{(\phi/2)} + \sin{(\psi/2)} \sin{(\theta/2)} \sin{(\phi/2)}
\tag{4.47a}
$$

$$
q_1 = \cos{(\psi/2)} \cos{(\theta/2)} \sin{(\phi/2)} - \sin{(\psi/2)} \sin{(\theta/2)} \cos{(\phi/2)}
\tag{4.47b}
$$

$$
q_2 = \cos{(\psi/2)} \sin{(\theta/2)} \cos{(\phi/2)} + \sin{(\psi/2)} \cos{(\theta/2)} \sin{(\phi/2)}
\tag{4.47c}
$$

and

$$
q_3 = \sin{(\psi/2)} \cos{(\theta/2)} \cos{(\phi/2)} - \cos{(\psi/2)} \sin{(\theta/2)} \sin{(\phi/2)}
\tag{4.47d}
$$

It can be shown that the quaternion vector obeys the differential equation

$$
\dot{q} = A(t)q
\tag{4.48}
$$

where

$$
A(t) = \frac{1}{2} \begin{bmatrix} 0 & -\Omega_z & -\Omega_y & -\Omega_x \\ \Omega_z & 0 & -\Omega_x & \Omega_y \\ \Omega_y & \Omega_x & 0 & -\Omega_z \\ \Omega_x & -\Omega_y & \Omega_z & 0 \end{bmatrix}
\tag{4.49}
$$

The entries in the matrix $A(t)$, Ω_x, Ω_y, and Ω_z are seen to be the computed body angular rates. The rates, Ω_x and Ω_y, are interchanged here as compared to where they would appear with the frame definition used by those in the aerospace field. This is because pitch is about the x axis here versus about the y axis there and vice versa for roll.

This differential equation for q can be integrated numerically to yield its updated values. If one uses the approximation that the matrix $A(t)$ is constant during the time interval from kT to $(k + 1)T$ with value $A(kT)$, then the solution would be

$$
q((k+1)T) = e^{A(kT)T} q(kT)
$$

If one were to integrate the yaw, pitch, and roll equations directly, the expressions for their derivatives in terms of body rates are more complex as was illustrated by equation (4.45). After integrating the quaternions the current values for yaw, pitch, and roll angles can be determined from the entries of q via

$$\psi = \tan^{-1}\left[\frac{2q_1q_2 + 2q_0q_3}{1 - 2q_2 - 2q_3}\right] \tag{4.50a}$$

$$\theta = -\sin^{-1}[2q_1q_3 - 2q_0q_2] \tag{4.50b}$$

and

$$\phi = \tan^{-1}\left[\frac{2q_2q_3 + 2q_0q_1}{1 - 2q_1 - 2q_2}\right] \tag{4.50c}$$

where use has been made of the fact that

$$q_0^2 + q_1^2 + q_2^2 + q_3^2 = 1$$

These new values can then be used for updating the rotation matrix. The quaternion representation has a number of desirable properties. One interesting feature is that the A matrix is skew symmetric, i.e.,

$$A = -A^T$$

As a result of this the square of the norm of the quaternion is seen to be constant

$$\frac{d}{dt}\left(q^Tq\right) = \dot{q}^Tq + q^T\dot{q}$$

or

$$\frac{d}{dt}\left(q^Tq\right) = q^TA^Tq + q^TAq$$

which is seen to be

$$\frac{d}{dt}\left(q^Tq\right) = q^T\left(A^T + A\right)q = 0$$

This result is consistent with the earlier mentioned fact that the square of the Euclidean norm is fixed at unity. This property can be used as a means to check and correct some of the numerical errors that accompany the numerical integration.

Another benefit from using the quaternions is that the singularity at pitch of 90° is avoided. Consider a rotation which consists of a yaw of amount ψ, followed by a pitch of $\pi/2$, followed by a roll of amount ϕ. This same final attitude could be attained by an infinite number of other combinations of yaw and roll as long as the sum of yaw plus roll remains the same. Thus it is impossible to determine a unique set of rotations corresponding to the final attitude. This situation is referred to as a singularity. This may also be detected from equation (4.36) by noting that there is division by $\cos\theta$ whose result is undefined for $\theta = \pm\pi/2$. The four components of the quaternion provide the minimum dimension necessary to avoid the singularity and yield a unique attitude even when pitch is 90°.

Regarding position calculation, the fact that the platform is rigidly attached to the vehicle means that the accelerometers measure acceleration in vehicle coordinates. These must be converted to inertial coordinates in order to be useful for determining the vehicle position. Thus one uses the rotation matrices to perform the transformation. Obviously the rotation matrices utilize the most recent computations of the vehicle orientation angles, ψ, θ, and ϕ.

$$\begin{bmatrix} a_x \\ a_y \\ a_z \end{bmatrix}_{inertial\ coords} = \begin{bmatrix} & Rot(\psi) & \end{bmatrix}\begin{bmatrix} & Rot(\theta) & \end{bmatrix}\begin{bmatrix} & Rot(\phi) & \end{bmatrix}\begin{bmatrix} a_x \\ a_y \\ a_z \end{bmatrix}_{vehicle\ coords}$$

$$(4.51)$$

One then performs a double integration of these accelerations as before to determine the new vehicle position. The equations above must all be executed at each sample instant. It is clear that the equations are much more complex for the case of strap-down gyros than for gimbaled gyros. However, construction costs for gimbaled gyros can be very high. Complex mechanical hardware for the gimbaled gyro is replaced by complex software operations for the strap-down gyro.

The ring laser gyro is another type of strap-down gyroscope. It measures the time it takes for light to make a complete circuit through a glass enclosure or through a coil of optical fiber. By sending two light beams in opposite directions, the difference in arrival time of the two beams can be used to compute the angular motion and thus the angular body rate about that axis. Again, quaternions may be used as a means to keep track of the attitude of the vehicle given the angular body rates. Since no gimbals are involved, translational accelerations are measured in body coordinates and must be converted to inertial coordinates before integration to obtain velocity and position calculation. Ring laser gyros are quite accurate for attitude. Translational position and velocity computations are still at the mercy of accelerometer bias. Computations and software requirements for these two types of strap-down gyros are comparable.

Low-cost micro-electromechanical-systems devices provide another means of measuring attitude. These devices use vibrational motion of solid-state devices, and the sensed Coriolis forces are used to compute attitude change. While the simplicity of construction and low cost make these very attractive for certain situations, limits on accuracy and precision may preclude their use in some applications.

Some of the navigation errors that may result from inertial systems are listed below.

Instrumentation errors: The sensed quantities may not exactly equal the physical quantities because of imperfections in the sensors (e.g., bias, scale factor, nonlinearity, random noise).

Computational errors: The navigation equations are typically implemented by a digital computer. Imperfect solutions of differential equations as a result of having approximated them as difference equations comprise another source of error.

Alignment errors: Errors caused by the fact that the sensors and their platforms may not be aligned perfectly with their assumed directions.

4.11 Dead Reckoning or Deduced Reckoning

Dead Reckoning typically uses shaft encoders or similar devices to measure the angular rotation of the wheels. The simple formula that follows is then used to convert this measurement to distance traveled.

$$AS = r\Delta\theta \tag{4.52}$$

In the above, r is the wheel radius. This yields length of the path traveled by the wheel, but it does not result in the new position since it does not contain any information about the curvature of the path traveled. To determine this, direction must also be recorded. One way to accomplish this is to use two encoders. Placing an encoder on each rear wheel works for front wheel steered vehicles or for independent wheel drive vehicles. Using a profile of each encoder reading, one can track vehicle motion in terms of direction and distance traveled and can compute its new position given its initial position. The following equations give the incremental changes in x position, y position, and heading.

$$\Delta\psi = \frac{r(\Delta\theta_R - \Delta\theta_l)}{W} \tag{4.53a}$$

$$\Delta x = -\frac{r(\Delta\theta_R + \Delta\theta_l)}{2}\sin\psi \tag{4.53b}$$

$$\Delta y = \frac{r(\Delta\theta_R + \Delta\theta_l)}{2}\cos\psi \tag{4.53c}$$

Here W is the lateral distance between the wheels, r is the wheel radius, and the $\Delta\theta$'s are the incremental encoder readings expressed in radians. These can be expressed as difference equations

$$\psi(k+1) = \psi(k) + \frac{r}{W}([\theta_r(k+1) - \theta_r(k)] - [\theta_l(k+1) - \theta_l(k)]) \tag{4.54a}$$

$$x(k+1) = x(k) - \frac{r}{2}\left([\theta_r(k+1) - \theta_r(k)] + [\theta_l(k+1) - \theta_l(k)]\right)\sin\left(\psi(k+1)\right)$$

$$(4.54b)$$

and

$$y(k+1) = y(k) + \frac{r}{2}\left([\theta_r(k+1) - \theta_r(k)] + [\theta_l(k+1) - \theta_l(k)]\right)\cos\left(\psi(k+1)\right)$$

$$(4.54c)$$

It should be apparent that a little wheel slippage can cause large error build-ups. For example, a slight error in heading can cause a large error in calculated location if the distance traveled is great. Thus this method of deduced reckoning, sometimes called "dead reckoning," can only be used for short distances and needs frequent re-calibration. Potential applications are in situations where contact with GPS satellites has been temporarily lost.

Example 9 *Let the wheel radius be 0.15 m and the lateral distance between tires 1 m. Suppose that $\psi(0) = -\pi/2$, $x(0) = 0$, and $y(0) = 0$. Let $\theta_l(k) = 0$, $k \le 15$, and $\theta_l(k) = (k-15)$, $16 < k < 40$. Let $\theta_r(k) = k$, $0 < k \le 30$ and $\theta_r(k) = 30$, $31 < k \le 40$. Compute the trajectory and plot $y(k)$ versus $x(k)$.*

Solution 9

The trajectory is generated using the difference equations for x, y, and heading, equation (4.45). The result is shown in Figure 4.7.

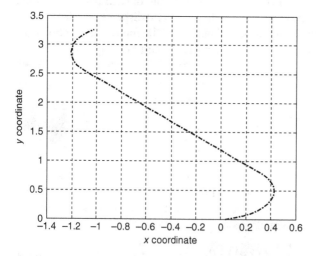

Figure 4.7 Robot trajectory based on dead reckoning, y versus x.

4.12   Inclinometer/Compass

The Inclinometer–Compass measures the rotation of the longitudinal axis about the original z axis, i.e., yaw, via a digital compass. It measures the angle of the longitudinal axis with respect to the original xy plane, i.e., pitch, via the gravity vector. It measures the rotation of the body about its longitudinal axis, i.e., roll, also via the gravity vector.

For the determination of pitch and roll from the sensed gravitational force, one may use the general rotation matrix to write

$$
\begin{bmatrix} g_x \\ g_y \\ g_z \end{bmatrix}_1 = \begin{bmatrix} \cos\psi\,\cos\phi - \sin\psi\,\sin\theta\,\sin\phi & -\sin\psi\,\cos\theta & \cos\psi\,\sin\phi + \sin\psi\,\sin\theta\,\cos\phi \\ \sin\psi\,\cos\phi + \cos\psi\,\sin\theta\,\sin\phi & \cos\psi\,\cos\theta & \sin\psi\,\sin\phi - \cos\psi\,\sin\theta\,\cos\phi \\ -\cos\theta\,\sin\phi & \sin\theta & \cos\theta\,\cos\phi \end{bmatrix}_{21} \begin{bmatrix} g_x \\ g_y \\ g_z \end{bmatrix}_2
$$

Here the force vector is measured along the respective axes of the rotated frame. The rotation matrix converts these forces to the original frame for which the z axis is vertical. In the original frame the gravitational force is zero except along the z axis, permitting one to write

$$
\begin{bmatrix} 0 \\ 0 \\ -g \end{bmatrix}_1 = \begin{bmatrix} \cos\psi\,\cos\phi - \sin\psi\,\sin\theta\,\sin\phi & -\sin\psi\,\cos\theta & \cos\psi\,\sin\phi + \sin\psi\,\sin\theta\,\cos\phi \\ \sin\psi\,\cos\phi + \cos\psi\,\sin\theta\,\sin\phi & \cos\psi\,\cos\theta & \sin\psi\,\sin\phi - \cos\psi\,\sin\theta\,\cos\phi \\ -\cos\theta\,\sin\phi & \sin\theta & \cos\theta\,\cos\phi \end{bmatrix}_{21} \begin{bmatrix} g_x \\ g_y \\ g_z \end{bmatrix}_2
$$

Now since the components of gravitational force along the various axes are independent of yaw, one may simplify the above relationships by considering the case where yaw is zero. Under this condition the above equation simplifies to

$$
\begin{bmatrix} 0 \\ 0 \\ -g \end{bmatrix}_1 = \begin{bmatrix} \cos\phi & 0 & \sin\phi \\ \sin\theta\,\sin\phi & \cos\theta & -\sin\theta\,\cos\phi \\ -\cos\theta\,\sin\phi & \sin\theta & \cos\theta\,\cos\phi \end{bmatrix}_{21} \begin{bmatrix} g_x \\ g_y \\ g_z \end{bmatrix}_2 \tag{4.55}
$$

From the first equation one obtains for roll

$$
\phi = \tan^{-1}\left(-g_x/g_z\right) \tag{4.56a}
$$

Now in the second equation one can make use of the first result and replace g_x by

$$
g_x = -g_z \sin\phi / \cos\phi
$$

After some algebra one then obtains for pitch

$$
\theta = \tan^{-1}\left(g_y \cos\phi / g_z\right) \tag{4.56b}
$$

Now the magnetic compass is used to determine yaw through the detection of the magnetic field along each axis. Consider a vehicle pointing North with zero pitch and roll. The magnetic field detected would be exclusively along the y axis. Now consider the vehicle yawed but still with zero pitch and roll. The magnetic field detected would be along both the x and y axes. Using the rotation matrix for yaw

$$R_{yaw}(\psi) = \begin{bmatrix} \cos\psi & -\sin\psi & 0 \\ \sin\psi & \cos\psi & 0 \\ 0 & 0 & 1 \end{bmatrix}$$

coupled with the above information yields

$$\begin{bmatrix} 0 \\ m_{y1} \\ 0 \end{bmatrix} = \begin{bmatrix} \cos\psi & -\sin\psi & 0 \\ \sin\psi & \cos\psi & 0 \\ 0 & 0 & 1 \end{bmatrix} \begin{bmatrix} m_{x2} \\ m_{y2} \\ m_{z2} \end{bmatrix}$$

or from the first equation

$$0 = m_{x2}\cos\psi - m_{y2}\sin\psi$$

with solution

$$\tan\psi = -m_{x2}/m_{y2}$$

This equation enables one to determine yaw from the components of the magnetic field detected along each axis. It is important to keep in mind that this result was derived under the assumption of zero pitch and roll. Next include pitch and roll as well as yaw. The relation between the vehicle with yaw alone and the one with yaw, pitch, and roll would be

$$\begin{bmatrix} \cos\psi & -\sin\psi & 0 \\ \sin\psi & \cos\psi & 0 \\ 0 & 0 & 1 \end{bmatrix} \begin{bmatrix} m_{x2} \\ m_{y2} \\ m_{z2} \end{bmatrix} = \begin{bmatrix} \cos\psi & -\sin\psi & 0 \\ \sin\psi & \cos\psi & 0 \\ 0 & 0 & 1 \end{bmatrix} \begin{bmatrix} 1 & 0 & 0 \\ 0 & \cos\theta & -\sin\theta \\ 0 & \sin\theta & \cos\theta \end{bmatrix}$$

$$\begin{bmatrix} \cos\phi & 0 & \sin\phi \\ 0 & 1 & 0 \\ -\sin\phi & 0 & \cos\phi \end{bmatrix} \begin{bmatrix} m_{x3} \\ m_{y3} \\ m_{z3} \end{bmatrix}$$

or

$$\begin{bmatrix} m_{x2} \\ m_{y2} \\ m_{z2} \end{bmatrix} = \begin{bmatrix} 1 & 0 & 0 \\ 0 & \cos\theta & -\sin\theta \\ 0 & \sin\theta & \cos\theta \end{bmatrix} \begin{bmatrix} \cos\phi & 0 & \sin\phi \\ 0 & 1 & 0 \\ -\sin\phi & 0 & \cos\phi \end{bmatrix} \begin{bmatrix} m_{x3} \\ m_{y3} \\ m_{z3} \end{bmatrix}$$

or

$$\begin{bmatrix} m_{x2} \\ m_{y2} \\ m_{z2} \end{bmatrix} = \begin{bmatrix} \cos\phi & 0 & \sin\phi \\ \sin\theta\sin\phi & \cos\theta & -\sin\theta\cos\phi \\ -\cos\theta\sin\phi & \sin\theta & \cos\theta\cos\phi \end{bmatrix} \begin{bmatrix} m_{x3} \\ m_{y3} \\ m_{z3} \end{bmatrix}$$

with solutions

$$m_{x2} = \cos\phi m_{x3} + \sin\phi m_{z3}$$

and

$$m_{y2} = \sin\theta\sin\phi m_{x3} + \cos\theta m_{y3} - \sin\theta\cos\phi m_{z3}$$

Finally one has

$$\tan\psi = -m_{x2}/m_{y2}$$
$$= -(\cos\phi m_{x3} + \sin\phi m_{z3})/(\sin\theta\sin\phi m_{x3} + \cos\theta m_{y3} - \sin\theta\cos\phi m_{z3})$$

$$(4.56c)$$

It is important to realize that the inclinometer responds to acceleration. If the vehicle is stationary or is moving in a straight line at a constant rate, the only acceleration is gravity, and the instrument provides a correct indication of pitch and roll. However, for any other case the indicated pitch and roll will be erroneous. Thus, some other means must be sought for dynamic situations. The inclinometer could be used in situations where the robot comes to a stop and then performs some action such as acquiring a radar image or infrared image. The attitude measurement from the Inclinometer/Compass could then be used to convert the image from robot coordinates to earth coordinates.

Exercises

1 A local coordinate system is set up at longitude of 77 °W and latitude of 38 °N. The x axis points East and the y axis points North. A mobile robot is determined to be located at 76.95 °W and 38.02 °N. Find the robot coordinates in the local frame in meters. Assume a spherical earth with radius of 6,378,137 m.

2 For a given *long* = 85 °W and *lat* = 42 °N, find X, Y, Z in ECEF coordinates. Use dimensions of meters and assume the point is on earth's surface. Take as the earth's radius 6,378,137 m.

3 Consider the point for which the ECEF coordinates are: $X = 3,000,000$, $Y = -5,000,000$, and $Z = -2,638,181$. Find *long* and *lat* (Note: The point may be slightly off the earth's surface).

4 Find UTM coordinates for the point *long* = 10 °E, *lat* = 43 °N.

5 A robot's GPS reads *long* = 85° and *lat* = 28°. Determine the longitudinal strip for UTM coordinates. Determine the central meridian for this strip. Using the spherical earth approximation determine the UTM coordinates, i.e., Easting and Northing.

6 A robot travels from the origin of a set of local coordinates (*long*$_0$ = −77° and *lat*$_0$ = 38°) to a point *long* = −77.01° and *lat* = 38.03°. Find the distance between the robot and the origin. Ignore the *z* component.

7 A robot on an exploration finds an object of interest. With respect to the robot, the object is 15 m in front and 3 m to the left. The robot's position in UTM coordinates is Easting = 246,315.2 and Northing = 1,432,765.4. Its heading is 60° measured counter-clockwise from East. Both pitch and roll are zero. What are the UTM coordinates of the object of interest?

8 A DGPS system is being used for more precise location determination. The base receiver is placed at a surveyed location with Easting = 276,453.12 and Northing = 1,235,462.76. After traveling some distance the robot reads as its location Easting = 276,602.32 and Northing = 1,235,813.76 while the base station reads as its location Easting = 276,454.62 and Northing = 1,235,464.15. What is the corrected robot location?

9 An array of four GPS antennas has the following respective position vectors in the order of Easting (*x*), Northing (*y*), and elevation (*z*). Both *L* and *W* are 1 m.

$$v1 = 1.0e + 006 \times 0.65342746650635$$
$$1.0e + 006 \times 1.34527600549365$$
$$1.0e + 006 \times 0.00003969640952$$

$$v2 = 1.0e + 006 \times 0.65342745390796$$
$$1.0e + 006 \times 1.34527671161197$$
$$1.0e + 006 \times 0.00003966122139$$

$$v3 = 1.0e + 006 \times 0.65342678349365$$
$$1.0e + 006 \times 1.34527668850635$$
$$1.0e + 006 \times 0.00003943759048$$

$$v4 = 1.0e + 006 \times 0.65342679609204$$
$$1.0e + 006 \times 1.34527598238804$$
$$1.0e + 006 \times 0.00003947277861$$

Find the robot attitude in terms of yaw, pitch, and roll. (See convention for receiver placements in Chapter 5.)

At a given time, a set of satellite positions in ECEF coordinates were given as

$x1 = 7,766,188.44$

$y1 = -21,960,535.34$

$z1 = 12,522,838.56$

$x2 = -25,922,679.66$

$y2 = -6,629,461.28$

$z2 = 31,864.37$

$x3 = -5,743,774.02$

$y3 = -25,828,319.92$

$z3 = 1,692,757.72$

$x4 = -2,786,005.69$

$y4 = -15,900,725.8$

$z4 = 21,302,003.49$

At the corresponding time, the ranges from a receiver are measured to be

$d1 = 1,022,228,206.42$

$d2 = 1,024,096,139.11$

$d3 = 1,021,729,070.63$

$d4 = 1,021,259,581.09$

Compute the location of the receiver. Work in ECEF coordinates. Since data from four satellites are provided, you have enough equations to solve not only for the location, but also determine the local clock error. You may use the following equations

$$d^1 = \left[\left(X^1 - x\right)^2 + \left(Y^1 - y\right)^2 + \left(Z^1 - z\right)^2\right]^{0.5} + ctr$$

$$d^2 = \left[\left(X^2 - x\right)^2 + \left(Y^2 - y\right)^2 + \left(Z^2 - z\right)^2\right]^{0.5} + ctr$$

$$d^3 = \left[\left(X^3 - x\right)^2 + \left(Y^3 - y\right)^2 + \left(Z^4 - z\right)^2\right]^{0.5} + ctr$$

$$d^4 = \left[\left(X^4 - x\right)^2 + \left(Y^4 - y\right)^2 + \left(Z^4 - z\right)^2\right]^{0.5} + ctr$$

where the d^i are the measured pseudo-ranges, (X^i, Y^i, Z^i) are the ECEF coordinates of iteratively for x, y, and z. Several iterations may be required to determine the values of x, satellite i, (x, y, z) are the ECEF position coordinates of the receiver antenna, where tr is the receiver clock bias and c is the speed of light. Solve these nonlinear equations y, and z that are consistent with the measured d^is. Also determine the local clock error.

A differential-wheel steered robot uses deduced reckoning to keep track of its position. Each wheel has radius of 0.2 m. The width of the robot is 1 m. The initial heading is zero, and the initial position is 0, 0. For the right side the rotation of the wheel is measured via an encoder and is given by,

$$\theta_{right}(k) = 0.5*k$$

For the left side it is

$$\theta_{left}(k) = 0.0005*k^3 - 0.01*k^2 + 0.5*k$$

Plot x versus k, y versus k, and y versus x for $0 \le k \le 20$.

At a given point signals are received from four different satellites. At the time the signal left the satellites their positions expressed in ECEF coordinates were as follows:

$x1 = 7,766,188.44$

$y1 = -21,960,535.34$

$z1 = 12,522,838.56$

$x2 = -25,922,679.66$

$y2 = -6,629,461.28$

$z2 = 31,864.37$

$x3 = -5,743,774.02$

$y3 = -25,828,319.92$

$z3 = 1,692,757.72$

$x4 = -2,786,005.69$

$y4 = -15,900,725.8$

$z4 = 21,302,003.49$

Based on the times the signals were received and the information regarding when they were transmitted from the respective satellites, their distances from the receiver are computed to be

$d1 = 1,022,228,210.42$

$d2 = 1,024,096,127.11$

$d3 = 1,021,729,065.63$

$d4 = 1,021,259,591.09$

What is the receiver location? Do not worry if the point is not on the earth's surface. What is the local clock error?

Consider the four antenna GPS array of a different configuration than the one analyzed in the text. Let the coordinates of each antenna in the vehicle frame be given as follows: $x1 = W/2, y1 = L/2, z1 = 0, x2 = W/2, y2 = -L/2, z2 = 0, x3 = -W/2, y3 = -L/2, z3 = 0$, and $x4 = -W/2, y4 = L/2, z4 = 0$. The position of each antenna is provided via the GPS receiver. Use the transformation matrix for yaw, pitch, and roll in that order to compute the position of each corner of the array for arbitrary attitude. Decide how to combine x, y, or z measurements of the different corners to isolate some convenient trigonometric function of yaw and solve for yaw in terms of these measurements $x1 = W/2, y1 = L/2, z1 = 0$. Do the same for pitch and roll. Use your own judgment in deciding what combinations are less susceptible to error. Your final answer should be a set of expressions for yaw, pitch, and roll in terms of the measured positions of the array corners.

References

Aviation Theory: Global Positioning System, http://www.flightsimaviation.com/aviation_theory_9_Global_Positioning_System_GPS.html (accessed September 13, 2019).

Brogan, W. L., *Modern Control Theory*, Prentice Hall, Upper Saddle River, NJ, 1991. http://www.colorado.edu/geography/gcraft/notes/coordsys/coordsys.html (accessed May 30, 2011).

Dana, P. H., *The Global Positioning System*, 2000. http://www.colorado.edu/geography/gcraft/notes/gps/gps_f.html. (accessed May 30, 2011).

El-Rabbany, A., *Introduction to GPS the Global Positioning System*, Artech House, 2002.

Farrel, Jay A. and Barth, Mathew, *The Global Positioning System & Inertial Navigation*, McGraw-Hill Professional, 1998.

GPS Primer Elements of GPS, http://www.aero.org/education/primers/gps/elements.html (accessed May 30, 2011).

Johnny Appleseed GPS – The Theory and Practice of GPS, http://www.ja-gps.com.au/what-is-gps.aspx (accessed September 13, 2019).

Kuipers, J. B., *Quaternions and Rotation Sequences*, Princeton University Press, Princeton, NJ, 1999.

National Geospatial-Intelligence Agency, *Military Map Reading 201 (PDF)*, May 29, 2002.

Widrow, B. and Stearns, S.D., *Adaptive Signal Processing*, Prentice Hall, 1985.

Yen, Kenneth and Cook, Gerald, "Improved Local Linearization Algorithm for Solving the Quaternion Equations," *AIAA Journal of Guidance and Control*, Vol. **3**, No. 5 (1980), pp. 468–471.

5

Application of Kalman Filtering

5.1 Introduction

This chapter is devoted to the estimation of nondeterministic quantities with special focus being on estimation of the state of a dynamic system, e.g., the location and orientation of a mobile robot, and also estimation of the coordinates of a detected object of interest. The Kalman Filter is presented and utilized to a large extent. Simulations are used to illustrate the capability of this methodology.

5.2 Estimating a Fixed Quantity Using Batch Processing

The accuracy of the geolocation of objects of interest depends on many factors, not the least of which is the accuracy of the estimates of the robot's position and attitude. In order to enhance this accuracy, a Kalman Filter may be utilized in conjunction with the measurements and the kinematic model. The development of the Kalman Filter will begin with a well-known estimation problem, that of estimating a fixed quantity using batch processing.

First define the vectors

$$Y_k = \begin{bmatrix} y_1 \\ y_2 \\ y_k \end{bmatrix} \quad (5.1)$$

and

$$V_k = \begin{bmatrix} v_1 \\ v_2 \\ v_k \end{bmatrix} \quad (5.2)$$

Mobile Robots: Navigation, Control and Sensing, Surface Robots and AUVs,
Second Edition. Gerald Cook and Feitian Zhang.
© 2020 by The Institute of Electrical and Electronics Engineers, Inc.
Published 2020 by John Wiley & Sons, Inc.

We have a set of measurement equations with error V_k relating the quantity to be estimated, x, to the measurements, Y_k.

$$Y_k = Hx + V_k \tag{5.3}$$

The errors are assumed to have zero mean and covariance described by

$$\text{cov}(V_k) = R \tag{5.4}$$

We form a function to be minimized utilizing the inverse of the covariance as the weighting matrix

$$J = (Y_k - H\hat{x}_k)^T R^{-1} (Y_k - H\hat{x}_k) \tag{5.5}$$

We seek to find $\hat{x}_k$, which minimizes this performance measure. By using the inverse of the covariance as the weighting matrix one causes those measurements with smallest measurement error covariance to be emphasized most heavily. The solution to this least-squares problem is well known to be

$$\hat{x}_k = \left[H^T R^{-1} H \right]^{-1} H^T R^{-1} Y_k \tag{5.6}$$

The covariance of the error in this estimate may be determined by first replacing Y_k in the above equation utilizing equation (5.3). Doing this yields

$$x - \hat{x}_k = x - \left[H^T R^{-1} H \right]^{-1} H^T R^{-1} (Hx + V_k)$$

or

$$x - \hat{x}_k = - \left[H^T R^{-1} H \right]^{-1} H^T R^{-1} V_k$$

Then forming the expectation of the outer product, recognizing that the expected value of the error itself is zero, yields for the covariance

$$E\left\{ (x - \hat{x}_k)(x - \hat{x}_k)^T \right\} = \left[H^T R^{-1} H \right]^{-1}$$

which is denoted as P_k. Thus

$$P_k = \left[H^T R^{-1} H \right]^{-1} \tag{5.7}$$

5.3   Estimating a Fixed Quantity Using Recursive Processing

The above result is a weighted generalized inverse. Now, if additional data become available, it is desirable to update the estimate by incorporating the new data, but without reprocessing the data already accounted for, i.e., we do not wish to repeat the processing of the entire batch. Thus a recursive scheme is sought which will incorporate the new data by using it to bring about a correction to the previous estimate. The following development follows very closely in Sage and Melsa, chapter 6, section 7, and also in Brogan, chapter 6, section 8. The additional measurement obtained is given by

$$y_{k+1} = H_{k+1}x + v_{k+1} \tag{5.8}$$

As before we form a function to be minimized

$$J = \left[(Y_k - H\hat{x}_{k+1})^T \ (y_{k+1} - H_{k+1}\hat{x}_{k+1})^T \right] \begin{bmatrix} R^{-1} & 0 \\ 0 & R_{k+1}^{-1} \end{bmatrix} \begin{bmatrix} Y_k - H\hat{x}_{k+1} \\ y_{k+1} - H_{k+1}\hat{x}_{k+1} \end{bmatrix} \tag{5.9}$$

with solution given by

$$\hat{x}_{k+1} = \left\{ \begin{bmatrix} H^T & H_{k+1}^T \end{bmatrix} \begin{bmatrix} R^{-1} & 0 \\ 0 & R_{k+1}^{-1} \end{bmatrix} \begin{bmatrix} H \\ H_{k+1} \end{bmatrix} \right\}^{-1} \begin{bmatrix} H^T & H_{k+1}^T \end{bmatrix} \begin{bmatrix} R^{-1} & 0 \\ 0 & R_{k+1}^{-1} \end{bmatrix} \begin{bmatrix} Y_k \\ y_{k+1} \end{bmatrix} \tag{5.10}$$

or

$$\hat{x}_{k+1} = \left\{ H^T R^{-1} H + H_{k+1}^T R_{k+1}^{-1} H_{k+1} \right\}^{-1} \left[H^T R^{-1} Y_k + H_{k+1}^T R_{k+1}^{-1} y_{k+1} \right]$$

Now using the previously derived result given by the rearranged equation (5.7)

$$P_k^{-1} = H^T R^{-1} H$$

then

$$\hat{x}_{k+1} = \left\{ P_k^{-1} + H_{k+1}^T R_{k+1}^{-1} H_{k+1} \right\}^{-1} \left[H^T R^{-1} Y_k + H_{k+1}^T R_{k+1}^{-1} y_{k+1} \right]$$

or, through use of the Matrix Inversion Lemma,

$$\hat{x}_{k+1} = \left\{ P_k - P_k H_{k+1}^T \left[H_{k+1} P_k H_{k+1}^T + R_{k+1} \right]^{-1} H_{k+1} P_k \right\} \left[H^T R^{-1} Y_k + H_{k+1}^T R_{k+1}^{-1} y_{k+1} \right]$$

which can be written as

$$\hat{x}_{k+1} = \left[P_k H^T R^{-1} Y_k \right] - P_k H_{k+1}^T \left[H_{k+1} P_k H_{k+1}^T + R_{k+1} \right]^{-1} H_{k+1} \left[P_k H^T R^{-1} Y_k \right]$$

$$+ P_k H_{k+1}^T R_{k+1}^{-1} y_{k+1} - P_k H_{k+1}^T \left[H_{k+1} P_k H_{k+1}^T + R_{k+1} \right]^{-1} H_{k+1} P_k H_{k+1}^T R_{k+1}^{-1} y_{k+1}$$

or

$$\hat{x}_{k+1} = \hat{x}_k - P_k H_{k+1}^T \left[H_{k+1} P_k H_{k+1}^T + R_{k+1} \right]^{-1} H_{k+1} \hat{x}_k + P_k H_{k+1}^T$$
$$\left\{ I - \left[H_{k+1} P_k H_{k+1}^T + R_{k+1} \right]^{-1} H_{k+1} P_k H_{k+1}^T \right\} R_{k+1}^{-1} y_{k+1}$$

Now expressing

$$I = \left[H_{k+1} P_k H_{k+1}^T + R_{k+1} \right]^{-1} \left[H_{k+1} P_k H_{k+1}^T + R_{k+1} \right]$$

we have

$$\hat{x}_{k+1} = \hat{x}_k - P_k H_{k+1}^T \left[H_{k+1} P_k H_{k+1}^T + R_{k+1} \right]^{-1} H_{k+1} \hat{x}_k$$

$$+ P_k H_{k+1}^T \left\{ \left[H_{k+1} P_k H_{k+1}^T + R_{k+1} \right]^{-1} \left[H_{k+1} P_k H_{k+1}^T + R_{k+1} \right] \right.$$

$$\left. - \left[H_{k+1} P_k H_{k+1}^T + R_{k+1} \right]^{-1} \left[H_{k+1} P_k H_{k+1}^T \right] \right\} R_{k+1}^{-1} y_{k+1}$$

or

$$\hat{x}_{k+1} = \hat{x}_k - P_k H_{k+1}^T \left[H_{k+1} P_k H_{k+1}^T + R_{k+1} \right]^{-1} H_{k+1} \hat{x}_k$$

$$+ P_k H_{k+1}^T \left[H_{k+1} P_k H_{k+1}^T + R_{k+1} \right]^{-1}$$

$$\left[H_{k+1} P_k H_{k+1}^T + R_{k+1} - H_{k+1} P_k H_{k+1}^T \right] R_{k+1}^{-1} y_{k+1}$$

This reduces to

$$\hat{x}_{k+1} = \hat{x}_k - P_k H_{k+1}^T \left[H_{k+1} P_k H_{k+1}^T + R_{k+1} \right]^{-1} H_{k+1} \hat{x}_k + P_k H_{k+1}^T$$

$$\left[H_{k+1} P_k H_{k+1}^T + R_{k+1} \right]^{-1} R_{k+1} R_{k+1}^{-1} y_{k+1}$$

or

$$\hat{x}_{k+1} = \hat{x}_k + P_k H_{k+1}^T \left[H_{k+1} P_k H_{k+1}^T + R_{k+1} \right]^{-1} \left[y_{k+1} - H_{k+1} \hat{x}_k \right]$$

or finally

$$\hat{x}_{k+1} = \hat{x}_k + K_{k+1} \left[y_{k+1} - H_{k+1} \hat{x}_k \right] \tag{5.11}$$

where the definition has been made

$$K_{k+1} = P_k H_{k+1}^T \left[H_{k+1} P_k H_{k+1}^T + R_{k+1} \right]^{-1} \tag{5.12}$$

We see that the new estimate is given by the old estimate plus a correction. The correction consists of a residual, i.e., the difference between the new measurement and the expected new measurement based on the old estimate, multiplied by the optimal gain. This form is intuitively appealing and provides the efficiency of being recursive, i.e., it builds on the estimate up to this point in time and does not reprocess measurements that have already been processed. For additional recursive steps, recalling the definition for P_k one replaces H by

$$\begin{bmatrix} H \\ H_{k+1} \end{bmatrix} \text{ and } R^{-1} \text{ by } \begin{bmatrix} R^{-1} & 0 \\ 0 & R_{k+1}^{-1} \end{bmatrix} \text{ to obtain}$$

$$P_{k+1} = \left\{ \begin{bmatrix} H^T & H_{k+1}^T \end{bmatrix} \begin{bmatrix} R^{-1} & 0 \\ 0 & R_{k+1}^{-1} \end{bmatrix} \begin{bmatrix} H \\ H_{k+1} \end{bmatrix} \right\}^{-1} = \left[H^T R^{-1} H + H_{k+1}^T R_{k+1}^{-1} H_{k+1} \right]^{-1}$$

or

$$P_{k+1} = \left[P_k^{-1} + H_{k+1}^T H_{k+1}^{-1} H_{k+1} \right]^{-1} \tag{5.13}$$

Using the matrix inversion lemma this can be expressed as

$$P_{k+1} = P_k - P_k H_{k+1}^T \left[H_{k+1} P_k H_{k+1}^T + R_{k+1} \right]^{-1} H_{k+1} P_k \tag{5.14}$$

which does not require the existence of P_k^{-1}. Note from equation (5.14) that the norm of P_{k+1} will be less than the norm of P_k, indicating the improvement in the estimate as a result of the new measurement. The next value for the gain matrix can now be found using equation (5.12) with the appropriate subscript updates.

Example 1   *Estimate the y coordinate of a stationary object with n equally reliable unbiased measurements.*

Solution 1

Using batch processing one would compute the expected value of this coordinate to be

$$\hat{x}_n = \frac{1}{n}\sum_1^n y_i$$

Upon receiving an additional measurement one could repeat the batch processing and obtain

$$\hat{x}_{n+1} = \frac{1}{n+1}\sum_1^{n+1} y_i$$

However, the above is seen to be

$$\hat{x}_{n+1} = \frac{1}{n+1}\sum_1^n y_i + \frac{1}{n+1}y_{n+1}$$

or

$$\hat{x}_{n+1} = \frac{n}{n+1}\left[\frac{1}{n}\sum_1^n y_i\right] + \frac{1}{n+1}y_{n+1}$$

or

$$\hat{x}_{n+1} = \left[1 - \frac{1}{n+1}\right]\hat{x}_n + \frac{1}{n+1}y_{n+1}$$

and finally

$$\hat{x}_{n+1} = \hat{x}_n + \frac{1}{n+1}(y_{n+1} - \hat{x}_n)$$

This can be viewed as the old estimate plus a gain times the residual where the residual is the difference between the new measurement and the expected value of the new measurement based on the old estimate.

Example 2   *The actual value of a quantity is 5.0. Use only noisy measurements of this quantity to estimate its value. The noise, which is additive, is Gaussian with mean of zero and variance of 0.04. Use a sequence of 50 measurements.*

Solution 2

The software code for solving this problem is given below and the results are shown in Figure 5.1.

```
%Estimation Example
R = 0.04;
P = R;
y(1) = 5 + sqrt(R) * randn;
xest(1) = y(1);

for i = 1:49
    y(i+1) = 5 + sqrt(R) * randn;
    P = 1 / ((1 / P) + (1 / R));
    K = P / (P + R);
    xest(i+1) = xest(i) + K * (y(i+1)-xest(i));
end

figure
plot(y)
hold on
plot(xest)
```

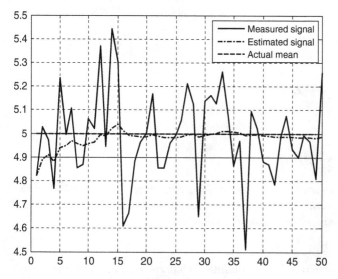

Figure 5.1 Sequence of measurements and estimates of the quantity, 5.0.

5.4 Estimating the State of a Dynamic System Recursively

The problem next considered differs from the problem addressed in the previous section, i.e., estimating a fixed quantity, in that it is seeks the estimate of the state of a dynamic system. Thus the state is changing from sample instant to sample instant. This problem is more complex than that of estimating a fixed quantity, and yet there is a relationship obeyed by this dynamic state that can be exploited in the estimation process. This relationship is the system equations.

Here we assume a linear time invariant system operating in discrete time with the following description. The state equations are

$$X((k+1)T) = AX(kT) + Bu(kT) + Gw(kT) \tag{5.15}$$

with output equation

$$Y((k+1)T) = HX((k+1)T) + v((k+1)T) \tag{5.16}$$

Here w represents a Gaussian, zero-mean, independent process disturbance of covariance Q and v represents Gaussian, zero-mean, independent measurement noise of covariance R. For shorthand notation, we shall represent the output at time $(k+1)T$, i.e., $Y((k+1)T)$ as Y_{k+1}. Denote the estimate of the state at time kT given measurements through time kT as $\hat{X}_{k/k}$ and the expected value of the state at time $(k+1)T$ given measurements through time kT as $\hat{X}_{k+1/k}$. This latter quantity is a prediction and is obtained by computing the conditional expectation of the next state taking advantage of the state equations. The result from simply using the right-hand side of the above equations is

$$E\{X((k+1)T)/Y(kT)\} = E\{AX(kT)/Y(kT)\} \\ + E\{Bu(kT)/Y(kT)\} + E\{Gw(kT)/Y(kT)\} \tag{5.17}$$

Using the shorthand notation coupled with the fact that the input signal is known and the expected value of the disturbance is zero, this reduces to

$$\hat{X}_{k+1/k} = A\hat{X}_{k/k} + Bu_k \tag{5.18}$$

The one-step prediction equation is seen to be simply a propagation of the discrete-time model, building on the estimate at the previous sample. With no additional measurements available, this would be the best estimate of the new system state. It is the estimate of the state at time $(k+1)T$ given measurements through time kT.

The covariance of the error in the prediction may be determined by first recognizing that the expected value of the prediction error is zero. Thus, this covariance reduces to

$$E\left\{\left(X((k+1)T)/Y(k)-\hat{X}_{k+1/k}\right)\left(X((k+1)T)/Y(k)-\hat{X}_{k+1/k}\right)^T\right\} =$$
$$E\left\{\left(AX(kT)/Y(kT)+Bu(kT)+Gw(kT)-A\hat{X}_{k/k}-Bu(kT)\right)\right.$$
$$\left.\left(AX(kT)/Y(kT)+Bu(kT)+Gw(kT)-A\hat{X}_{k/k}-Bu(kT)\right)^T\right\}$$

Now since the present value of the state is independent of the present value of the process disturbance, this reduces to

$$P(k+1/k) = AE\left\{\left(X(kT)-\hat{X}_{k/k}\right)\left(X(kT)-\hat{X}_{k/k}\right)^T\right\}A^T + GQG^T$$

or

$$P(k+1/k) = AP(k/k)A^T + GQG^T$$

where

$$P(k/k) = E\left\{\left(X(kT)-X_{k/k}\right)\left(X(kT)-X_{k/k}\right)^T\right\}$$

is the covariance of the error in the estimate at the kth step. Here, it was made use of the fact that the expected value of the error in the estimate is zero.

Now when a new measurement becomes available, the information contained within it can be incorporated to improve the estimate. After obtaining the measurement at time $(k+1)T$, the new estimate is as follows:

$$E\{X((k+1)T)/Y(k+1)\} = \hat{X}_{k+1/k} + E\left\{\left(X((k+1)T)-X_{k+1/k}\right)/\right.$$
$$\left.\left(Y_{k+1}-H\hat{X}_{k+1/k}\right)\right\} \tag{5.19}$$

Note that the equation above has been arranged in such a way that the quantities in the conditional expectation on the right side are each of zero mean. This becomes

$$\hat{X}_{k+1/k+1} = \hat{X}_{k+1/k} + P_{xy}P_{yy}^{-1}\left(Y_{k+1}-H\hat{X}_{k+1/k}\right) \tag{5.20}$$

where the covariances in the above equation are defined as

$$P_{xy} = E\left\{\left(X_{k+1}-\hat{X}_{k+1/k}\right)\left(Y_{k+1}-H\hat{X}_{k+1/k}\right)^T\right\} \tag{5.21}$$

and

$$P_{yy} = E\left\{\left(Y_{k+1}-H\hat{X}_{k+1/k}\right)\left(Y_{k+1}-H\hat{X}_{k+1/k}\right)^T\right\} \tag{5.22}$$

The preceding is based on the known basic result for Gaussian variables

$$E(x/y) = \bar{x} + P_{xy}P_{yy}^{-1}(y-\bar{y}) \tag{5.23}$$

For the covariance P_{xy} it may be seen from its definition that

$$E\left\{ \left(X_{k+1} - \hat{X}_{k+1/k}\right)\left(Y_{k+1} - H\hat{X}_{k+1/k}\right)^T \right\}$$

$$= E\left\{ \left(X_{k+1} - \hat{X}_{k+1/k}\right)\left(HX_{k+1} + v_{k+1} - H\hat{X}_{k+1/k}\right)^T \right\}$$

Now since v_{k+1} is of zero mean, and also since X_{k+1} and $\hat{X}_{k+1/k}$ are independent of v_{k+1}, this reduces to

$$P_{xy} = P(k+1/k)H^T$$

Likewise, for the covariance P_{yy} it may be noted that

$$E\left\{ \left(Y_{k+1} - H\hat{X}_{k+1/k}\right)\left(Y_{k+1} - H\hat{X}_{k+1/k}\right)\right\}^T$$

$$= E\left\{ \left(HX_{k+1} + v_{k+1} - H\hat{X}_{k+1/k}\right)\left(HX_{k+1} + v_{+1} - H\hat{X}_{k+1/k}\right)^T \right\}$$

or

$$P_{yy} = HE\left\{ \left(X_{k+1} - \hat{X}_{k+1/k}\right)\left(X_{k+1} - \hat{X}_{k+1/k}\right)^T \right\}H^T + E\left\{v_{k+1}v_{k+1}^T\right\} = HP(k+1/k)H^T + R$$

Thus,

$$P_{xy}P_{yy}^{-1} = P(k+1/k)H^T\left[HP(k+1/k)H^T + R\right]^{-1}$$

The estimation equation may be rewritten in the form

$$\hat{X}_{k+1/k+1} = \hat{X}_{k+1/k} + K(k+1)\left(Y_{k+1} - H\hat{X}_{k+1/k}\right) \tag{5.24}$$

The estimate comprises the prediction plus a correction. This correction is a residual term premultiplied by the gain matrix $K(k+1)$. The residual term comprises the actual measurement minus the predicted value of this measurement. It is seen to depend on the predicted state through the output matrix and is an $m \times 1$ vector, which is usually of lower dimension than the state which is $n \times 1$. The covariance of the error in the estimate may be evaluated. First the error is

$$X_{k+1} - \hat{X}_{k+1/k+1} = X_{k+1} - \hat{X}_{k+1/k} - K(k+1)\left(Y_{k+1} - H\hat{X}_{k+1/k}\right)$$

or

$$X_{k+1} - \hat{X}_{k+1/k+1} = X_{k+1} - \hat{X}_{k+1/k} - K(k+1)\left(HX_{k+1} + v_{k+1} - H\hat{X}_{k+1/k}\right)$$

Forming the covariance as the expected value of this times its transpose yields

$$P(k+1/k+1) = P(k+1/k) - P(k+1/k)H^TK^T - KHP(k+1/k) + K\left(HP(k+1/k)H^T + R\right)K^T$$

Utilizing the expression above for $P_{xy}P_{yy}^{-1}$ as the gain matrix and performing some manipulations yields

$$P(k+1/k+1) = P(k+1/k) - P(k+1/k)H^T \left(HP(k+1/k)H^T + R\right)^{-1} HP(k+1/k) \tag{5.25}$$

The gain $K(k+1)$, which will be of dimension $n \times m$, may be obtained sequentially by solving the equations

$$P(k+1/k) = AP(k/k)A^T + GQG^T \tag{5.26}$$

which is the covariance of the error in the prediction, and

$$K(k+1) = P_{xy}P_{yy}^{-1} = P(k+1/k)H^T \left[HP(k+1/k)H^T + R\right]^{-1} \tag{5.27}$$

The filter gain may be used in turn for the following equation

$$P(k+1/k+1) = [I - K(k+1)H]P(k+1/k) \tag{5.28}$$

which is the covariance of the error in the estimate. The last equation may be shown to be equivalent to equation (5.25). This filter is optimal in terms of providing the expected value of the state when the system equations and the output equations are linear, when the process disturbance and measurement noise have Gaussian distributions, and when they are both independent sequences.

Example 3 *Consider the discrete-time dynamic system*

$$x_1(k+1) = x_1(k) + Tx_2(k)x_2(k+1) = x_2(k) + Tu(k) + w(k)$$

where x_1 is position and x_2 is velocity. The measurement equation is

$$y(k+1) = x_1(k+1) + v(k+1)$$

Let the system initially be at rest and let the input be a pulse of unit amplitude and duration $10\,T$ where T is taken to be 0.2 seconds. Take variances $R = 0.006$ and $Q = 0.00005$. Simulate this system and implement a Kalman filter to estimate the state.

Solution 3

The software code for solving this problem is given below and the results are shown in Figure 5.2a and b.

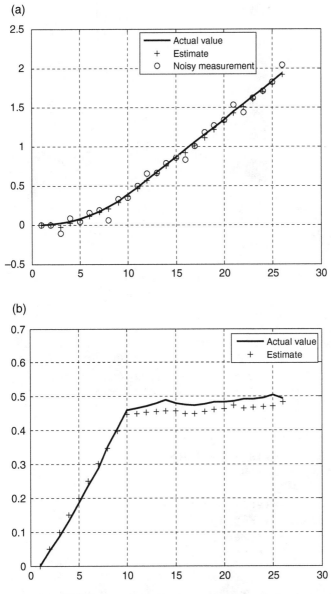

Figure 5.2 (a) Plot of actual value, noisy measurement, and estimate of x_1. (b) Plot of actual value and estimate of x_2. No measurement available.

```
T = 0.2;
A = [1 T; 0 1];
B = [T^2 / 2 T]';
H = [1 0];
G = [0 1]';
Q = 0.00005;
R = 0.006
x1(1) = 0;
x2(1) = 0;
x1e(1) = 0;
x2e(1) = 0;
xest = [x1e(1) x2e(1)]';
x1p(1) = 0;
x2p(1) = 0;
PE = [R 0; 0 0];
PP = A * PE(1) * A' + Q;

for i= 1:25
    if i < 10
       u = 0.25;
    else
       u = 0;
    end
    x1(i+1) = x1(i) + T * x2(i) + (T^2 / 2) * u;
    x2(i+1) = x2(i) + T * u + sqrt(Q) * randn;
    y(i+1) = x1(i+1) + sqrt(R) * randn;

    PP = A * PE * A' + G * Q * G';
    K = PP * H' * inv(H * PP * H' + R);
    PE = [eye(2) - K * H] * PP;

    xpredict = A * xest + B * u;
    xest = xpredict + K * (y(i+1) -H * xpredict);
    x1e(i+1) = [1 0] * xest;
    x2e(i+1) = [0 1] * xest;
end
```

The benefit of the filtering is quite apparent for the estimate of position when compared to the noisy measurement and especially apparent for the estimate of velocity that was not even measured.

It is interesting to further analyze the contributions of the two terms that make up the estimate of state. The prediction is the propagation of the previous estimate utilizing the system model. The reliability of this model is manifested through one's choice of Q, the covariance of the disturbance w. If one has confidence in the fidelity of the model, then the Q matrix is chosen to represent only the process disturbance. If one knows that the model is only a rough approximation, then the Q matrix is chosen to be larger to account for this uncertainty as well as the process disturbance. When this covariance is small, then the model can do a good job of predicting the next value of the state. However, if this covariance is large, then the prediction is not very reliable, and the output measurement needs to play a greater role in determining the next value of the state.

The reliability of the output measurement is manifested through one's choice of R, the covariance of the measurement noise, v. A small measurement noise covariance corresponds to an accurate output measurement with the result being that the residual correction is incorporated strongly into the estimate. Conversely, a large measurement noise covariance corresponds to an unreliable output measurement and the residual correction is given a lesser role in the estimate. These two covariances, Q and R, interact together to yield the appropriate weighting to the model propagation and to the residual correction.

The $P(k + 1/k)$ equation shows that the covariance of the error in prediction depends on the system equations through A and on the process disturbance. Compared to the covariance of the error in the estimate, this covariance of the error in prediction is always increased by the positive semi-definite disturbance term, GQG^T and may be increased or decreased by the $AP(k/k)A^T$ term depending on the eigenvalues of A.

To analyze the behavior of the covariance of the estimate, it is instructive to utilize the expression from equation (5.25) which is repeated here for convenience

$$P(k + 1/k + 1) = P(k + 1/k) - P(k + 1/k)H^T \left[HP(k + 1/k)H^T + R \right]^{-1} HP(k + 1/k)$$

$$(5.29)$$

The above reveals that the covariance of the error after the measurement has been taken, i.e., the covariance of the estimate, will always be smaller than the covariance of the error before the measurement was taken. This can be seen from the fact that any covariance matrix is positive semi-definite and that the term being subtracted off from $P(k + 1/k)$ is positive semi-definite. Thus $P(k + 1/k + 1)$ will be smaller in norm than $P(k + 1/k)$. This seems intuitively correct since the addition of the measurement can only provide more information as to the true value of the state.

Example 4 *Analyze the optimal filter for the simple scalar system*

$$x(k + 1) = ax(k) + u(k) + w(k)$$

with

$$y(k) = x(k) + v(k)$$

Solution 4

The filter equations are

$$\hat{x}_{k+1/k} = \alpha\hat{x}_{k/k} + u(k)$$

the prediction, and

$$\hat{x}_{k+1/k+1} = \hat{x}_{k+1/k} + K(k+1)\left(y_{k+1} - \hat{x}_{k+1/k}\right)$$

the estimate.

Combining the equations for the covariance of the prediction and the covariance of the estimate yields a single equation for the estimate covariance

$$P(k+1/k+1) = [I - K(k+1)H]\left[AP(k/k)A^T + GQG^T\right]$$

Setting the steady state of the estimation covariance to p and using the model parameters for the example yields the equation

$$p = (1-K)\left(\alpha^2 p + q\right)$$

Using the equation for the covariance of the prediction in the equation for the filter gain yields

$$K(k+1) = \left[AP(k/k)A^T + GQG^T\right]H^T\left[H\left(AP(k/k)A^T + GQG^T\right)H^T + R\right]^{-1}$$

or, for this example

$$K = \frac{\alpha^2 p + q}{\alpha^2 p + q + r}$$

It is seen from the above that for the special case $r = 0$, K is unity. The equation

$$\hat{x}_{k+1/k+1} = \hat{x}_{k+1/k} + K\left(y_{k+1} - \hat{x}_{k+1/k}\right)$$

becomes

$$\hat{x}_{k+1/k+1} = \hat{x}_{k+1/k} + \left(y_{k+1} - \hat{x}_{k+1/k}\right)$$

or

$$\hat{x}_{k+1/k+1} = y_{k+1}$$

That is, the best estimate is the output measurement itself since it is noise free. Here the ratio of measurement noise to process disturbance is

$$r/q = 0$$

Using the expression for K in the equation for p gives the equation

$$p = \frac{r(\alpha^2 p + q)}{\alpha^2 p + q + r}$$

Evaluating this for r = 0 reveals that p is zero which would be expected with zero measurement noise.

When we now consider the opposite case, i.e., a measurement noise that is very large, it is seen that p takes on a nonzero but finite value. Dividing numerator and denominator of the expression for p above by r yields

$$p = \frac{(\alpha^2 p + q)}{\alpha^2 p/r + q/r + 1}$$

whose solution for large r becomes

$$p = \frac{q}{1 - \alpha^2}$$

The gain becomes

$$K = \frac{q}{q + (1 - \alpha^2)r}$$

which approaches zero for large r. In this case the estimator approaches

$$\hat{x}_{k+1/k+1} = \hat{x}_{k+1/k} + 0\left(y_{k+1} - \hat{x}_{k+1/k}\right)$$

or

$$\hat{x}_{k+1/k+1} = \hat{x}_{k+1/k}$$

That is, the best estimate is the propagation of the model since the measurement is very noisy. Here the ratio of process disturbance to measurement noise is

$$q/r = 0$$

These two extreme cases in the example above illustrate the way the filter allocates weighting on the measurement versus weighting on the propagation of the model. The confidence in each depends on the respective measurement noise and process disturbance covariances.

By using the Kalman Filter, it is possible to obtain estimates having lower error variance than the measurement itself. For the example just considered, we set alpha to be 0.707 and r to be 1.0. The parameter q was varied from 0.1 to 5.0 and the covariance of the prediction error and the covariance of the estimation error were computed and plotted. These are shown in Figure 5.3. The covariance of the estimate is the smaller one. Note that its value is very small for small q and that it increases with q but is always lower than r. In fact, it asymptotically approaches r as q gets large.

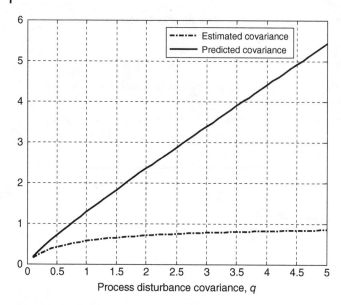

Figure 5.3 Plot of estimation covariance and prediction covariance versus process disturbance covariance, *q*. Measurement noise covariance, *r*, is 1.0.

It is also interesting to examine the value of gain under these same conditions. In Figure 5.4, it is seen that the gain is low when the ratio of process disturbance to measurement noise is small, indicating that the filter relies more on the model than on the relatively noisy measurement. When this ratio is large, the gain increases causing the filter to give more weight to the measurement. As *q* gets larger, the gain asymptotically approaches unity.

Conversely, it is interesting to observe the behavior of the prediction and estimate covariances when the covariance of the process disturbance is held constant and the covariance of the measurement noise is varied. As the measurement noise covariance get large, the estimate covariance and the prediction covariance both approach the same value, $q/(1 - \alpha^2)$; i.e., there is little or no additional improvement from using the very noisy measurement in the estimate versus the prediction. This is evident from the plot of estimate covariance and prediction covariance in Figure 5.5.

The behavior of the filter is also reflected in the plot of gain shown in Figure 5.6. When the covariance of the measurement noise is very large, the gain diminishes. The filter places its confidence in the propagation of the model and ignores the new measurement. It is important to keep in mind that these results shown in Figures 5.3–5.6 and in the preceding example, all apply only to the steady-state behavior of the filter.

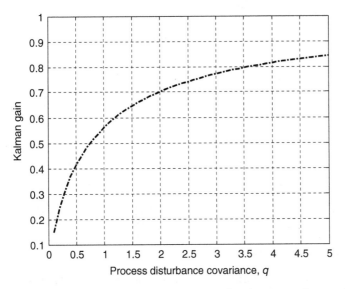

Figure 5.4 Plot of Kalman gain versus process disturbance covariance, q. Measurement noise covariance, r, is 1.0.

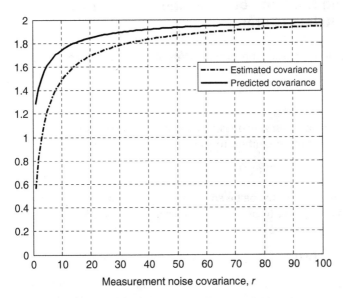

Figure 5.5 Plot of estimation covariance and prediction covariance versus measurement noise covariance, r. Process disturbance covariance, q, is 1.0.

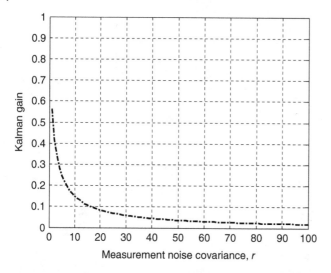

Figure 5.6 Plot of Kalman gain versus measurement noise covariance, r. Process disturbance covariance, q, is 1.0.

5.5 Estimating the State of a Nonlinear System via the Extended Kalman Filter

When the system is nonlinear and described in terms of a continuous-time model

$$\dot{X} = f(X, U) + Gw \tag{5.30}$$

where w represents a Gaussian zero-mean white-noise process disturbance, one can use Euler's method of integration to yield as a discrete-time model

$$X((k+1)T) = X(kT) + Tf(X(kT), U(kT)) + TGw(kT) \tag{5.31}$$

The sampling interval, T, must be taken sufficiently small in accordance with the system dynamics in order to provide an accurate representation of the system. The output equations are

$$Y((k+1)T) = h(X((k+1)T)) + v((k+1)T) \tag{5.32}$$

Here again v represents a random, zero-mean, white measurement noise. Again it is helpful to use shorthand notation and represent the output at time $(k+1)T$ as y_{k+1} and the estimate of the state at time kT given measurements through time kT as $\hat{X}_{k/k}$. The one-step prediction equation is simply a propagation of the discrete-time model given above and for this case is

$$E\{X((k+1)T)/Y_k\} = E\{(X(kT) + Tf(X(kT),U(kT)) + TGw(kT))/Y_k\}$$
(5.33)

or

$$\hat{X}_{k+1/k} = \hat{X}_{k/k} + TE\{f(X(kT),U(kT))\}$$
(5.34)

It is well known that in general

$$E\{f(X(kT),U(kT))/Y_k\} \neq f(\hat{X}_{k/k},U(kT))$$

To compute the solution to equation (5.34) would require knowledge of the probability density function $p(X, kT)$. In order to obtain a practical estimation algorithm, one may expand $f(X(kT), U(kT))$ via the Taylor series about the current estimate of the state vector through the linear term to obtain

$$f(X(kT),U(kT)) \approx f(\hat{X}_{k/k},U(kT)) + (\partial f/\partial X)|_{x=\hat{x}(k/k)}(X(kT)-\hat{X}_{k/k}) + \cdots$$
(5.35a)

and

$$E\{f(X(kT),U(kT))/Y_k\} \approx f(\hat{X}_{k/k},U(kT)) + (\partial f/\partial X)|_{x=\hat{x}(k/k)}E\{(X(kT)-\hat{X}_{k/k})/Y_k\} + \cdots$$
(5.35b)

or

$$E\{f(X(kT),U(kT))/Y_k\} \approx f(\hat{X}_{k/k},U(kT)) + 0 + \cdots$$
(5.35c)

Note that this approximation is correct through the linear term. For more discussion on this, see Gelb, A. (ed.), *Applied Optimal Estimation*, MIT Press, Cambridge, MA, 1974, pp. 183–184. Thus the prediction may be expressed using the shorthand notation as

$$\hat{X}_{k+1/k} = \hat{X}_{k/k} + Tf(\hat{X}_{k/k},U_k)$$
(5.36)

After obtaining the measurement at time $(k+1)T$, the new estimate is as follows

$$\hat{X}_{k+1/k+1} = \hat{X}_{k+1/k} + K(Y_{k+1} - h(\hat{X}_{k+1/k}))$$
(5.37)

i.e., the prediction plus a correction. The gain K is obtained by sequentially solving the same equations as before

$$P(k+1/k) = A(k)P(k/k)A(k)^T + (GT)Q(GT)^T$$
(5.38a)

$$K(k+1) = P(k+1/k)H(k)^T \left[H(k)P(k+1/k)H(k)^T + R\right]^{-1} \qquad (5.38b)$$

and

$$P(k+1/k+1) = [I - K(k+1)H(k)]P(k+1/k) \qquad (5.38c)$$

Here $A(k)$ is the linearization of the nonlinear discrete-time system equations at the present estimate of the state,

$$A(k) = I + T\left[\frac{\partial f(X,U)}{\partial X}\right]_{\hat{X}_{k/k},U_k} \qquad (5.39a)$$

and $H(k)$ is the linearization of the nonlinear output equation

$$H(k) = \left[\frac{\partial h(X)}{\partial X}\right]_{\hat{X}_{k/k}} \qquad (5.39b)$$

Also note from equation (5.38a) that the discrete-time equivalent of the disturbance coefficient matrix is GT. The matrix Q is the covariance of the process disturbance $w(kT)$, and the matrix R is the covariance of the measurement noise $v(kT)$. Here, as in the case of a linear system, we see the interaction between the covariance of the error in prediction, i.e., the error before the measurement is made, and the covariance of the error in the estimate, i.e., the error after the measurement is made. It is worth mentioning again that the covariance of the estimate error is always less than or equal to the covariance of the prediction error because more information has been made available.

If the nonlinear system is described initially in discrete time as

$$X_{k+1} = f(X_k, U_k) + Gw_k \qquad (5.40)$$

then the prediction step becomes simply

$$\hat{X}_{k+1/k} = f(\hat{X}_{k/k}, U_k) \qquad (5.41)$$

and the matrix $A(k)$ is defined by

$$A(k) = \left[\frac{\partial f(X_k, U_k)}{\partial X}\right]_{\hat{X}_{k/k},U_k} \qquad (5.42)$$

Further, there is no T factor multiplying the disturbance coefficient matrix in the equation for the covariance of the prediction error. Everything else remains the same.

Example 5 *Apply the Kalman Filter to a front-wheel steered mobile robot. The fifth-order model is to be used with state defined as*

$$X = \begin{bmatrix} x \\ y \\ \psi \\ v \\ \alpha \end{bmatrix}$$

The control algorithm for the steering is computed control. The desired closed-loop behavior is to have a natural frequency of 0.2 and a damping ratio of 1.0. Implementation of this steering algorithm requires steering angle, velocity, actual heading, and desired heading. The length of the robot is 1.0 m and the sample interval is taken to be 0.1 second. Only the x and y coordinates of the robot location are measured.

Solution 5

For the output, since it was assumed that only x and y were measured, we have

$$H = \begin{bmatrix} 1 & 0 & 0 & 0 & 0 \\ 0 & 1 & 0 & 0 & 0 \end{bmatrix}$$

Process disturbance was assumed in the heading and velocity equations. Thus,

$$G = \begin{bmatrix} 0 & 0 \\ 0 & 0 \\ 0 & 0 \\ 1 & 0 \\ 0 & 1 \end{bmatrix}$$

The covariance matrix for the measurement noises is

$$R = \begin{bmatrix} 0.04 & 0 \\ 0 & 0.04 \end{bmatrix}$$

For the disturbance the covariance matrix is

$$Q = \begin{bmatrix} 0.01 & 0 \\ 0 & 0.01 \end{bmatrix}$$

The matrix exponential for the discrete-time model is computed via the infinite series truncated after the fourth term. The other terms in the model follow accordingly.

$$\Delta = IT + A\frac{T^2}{2!} + A^2\frac{T^3}{3!} + A^3\frac{T^4}{4!}$$

$$A_{disc} = I + A*\Delta$$

and

$$G_{disc} = \Delta*G$$

The desired heading was taken to be the direction from the current location to the specified destination, $x = 25$, $y = 25$, and the desired velocity was taken to be $v = 0.5$ m/s. The simulation was halted before the final state had been reached.

In the Figure 5.7a, the displacement in the x coordinate is the variable of interest. The robot motion has been simulated; therefore, the actual value of x is available for the plot. Also plotted are the noisy measurement of x and the estimate of x. It is apparent that the estimate is not perfect; however, it is considerably better in accuracy than the measurement.

Figure 5.7b shows the actual, measured, and estimated motion in the y coordinate.

In Figure 5.7c, the path of the robot is shown in the x–y space.

It was assumed that there was no measurement of heading. However, the robot model used in the Kalman Filter contained heading as a state variable and thus leads to an estimate of it. This result points out that the filter not only improves the estimates of measured variables, but also provides estimates of unmeasured variables. Figure 5.7d shows the plots of heading.

It was also assumed that there was no measurement of velocity. Nevertheless the Kalman Filter provides an estimate of this quantity since it is an observable state. Plots of the actual and estimated velocity are shown in Figure 5.7e.

Steering angle was another component of state that was not measured. Here again the Kalman Filter provides an estimate of this quantity since it is an observable state. Plots of the actual and estimated steering angle are shown in Figure 5.7f.

In this example, the system was of order five and the number of outputs was two. Through the use of the Kalman Filter, estimates of all states were obtained. The steering and velocity control algorithms required knowledge of all the states; therefore, the estimates were used instead of the actual values. The preceding plots show that the robot is being steered toward the destination and that the velocity is being maintained near its desired value. This particular control algorithm, computed control for a specified natural frequency and damping ratio, stretches the limits of use for these nonperfect estimates of the states. A simpler control algorithm would be more robust. An alternative would be to add a digital compass to measure the heading.

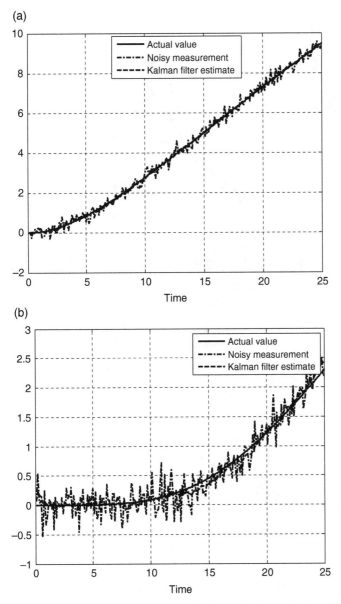

Figure 5.7 (a) *X* coordinate versus time for front-wheel steered robot. (b) *Y* coordinate versus time for front-wheel steered robot. (c) *Y* coordinate versus *X* coordinate for front-wheel steered robot. (d) Actual, estimated, and desired robot heading angle for front-wheel steered robot. No heading angle measurement was available. Estimates based on Kalman Filter with only *x* and *y* measurements. (e) Actual and estimated robot velocity for front-wheel steered robot. (f) Actual and estimated steering angle for front-wheel steered robot.

(c)

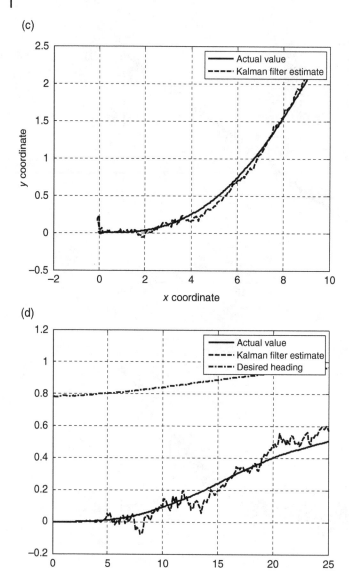

(d)

Figure 5.7 (Continued)

(e)

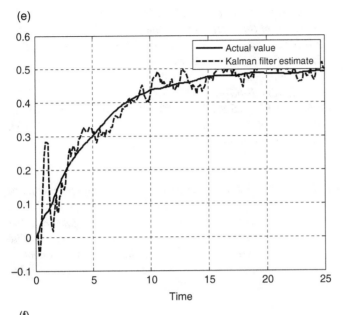

(f)

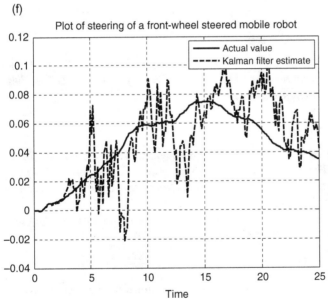

Figure 5.7 (Continued)

Example 6 *Apply the Kalman Filter to the GPS initialization problem, i.e., the determination of the location of the receiver given the pseudo distances from the visible satellites to the receiver and given the locations of these satellites. The data on the satellite positions in ECEF coordinates at time of transmit are as follows X1 = [7,766,188.44, −21,960,535.34, 12,522,838.56], X2 = [−25,922,679.66, −6,629,461.28, 31,864.37], X3 = [−5,743,774.02, −25,828,319.92, 1,692,757.72], and X4 = [−2,786,005.69, −15,900,725.8, 21,302,003.49] and the pseudo distances are given as d1 =1,022,228,206.42, d2 = 1,024,096,139.11, d3 = 1,021,729,070.63, and d4 =1,021,259,581.09, all in meters.*

Solution 6

The problem must be formulated to suit the format of an estimation problem. For the initialization, one essentially freezes time and processes the single set of measurements recursively until the estimate of receiver position converges. Thus, the positions of the satellites are treated as constant during this iterative process. The position of the receiver is also modeled as being constant but with the inclusion of a disturbance term to allow for its adjustment as the estimation process takes place. Recall that one has N equations of the form

$$d^i = \left\{ (X - X^i)(X - X^i)^T \right\}^{-0.5} - ct_b \text{ where the receiver coordinates are given by } X = [x, y, z]^T$$

Now define the N-dimensional output vector as $YY = \begin{bmatrix} d^1 & d^2 & d^3 & \dots & d^N \end{bmatrix}^T$ and the four-dimensional state vector as $XX = \begin{bmatrix} x & y & z & ctb \end{bmatrix}^T$. Using these variables, the output equation may be written as

$$YY(k) = h(XX(k)) + v(k)$$

Here the vector v represents the measurement noise, i.e., the uncertainty in the pseudo distances. These errors are caused by uncertainty in the correlation process of determining the time of arrival of signals from the satellites to the receiver and also by uncertainty in the satellites' positions.

Under the assumption that the receiver position is fixed and the local clock bias is constant the process model is simply

$$XX(k+1) = XX(k) + Iw$$

Here the 4 × 1 process disturbance vector X is included to allow for changes in X and ct_b as the estimation process takes place. Thus, we have

$$A(XX(k)) = I$$

and

$$G(k) = I$$

The output matrix is given by

$$H(XX(k)) = \partial h / \partial XX$$

which becomes the N × 4 matrix

$$H(XX) = \begin{bmatrix} \dfrac{x(k)-x^1}{r^1(k)} & \dfrac{y(k)-x^1}{r^1(k)} & \dfrac{z(k)-x^1}{r^1(k)} & -1 \\[2mm] \dfrac{x(k)-x^2}{r^1(k)} & \dfrac{y(k)-x^2}{r^1(k)} & \dfrac{z(k)-x^2}{r^1(k)} & -1 \\[2mm] \vdots & \vdots & \vdots & \vdots \\[2mm] \dfrac{x(k)-x^N}{r^1(k)} & \dfrac{y(k)-x^N}{r^1(k)} & \dfrac{z(k)-x^N}{r^1(k)} & -1 \end{bmatrix}.$$

R, the covariance matrix for the measurement noise is N × N and Q, the covariance matrix for the process disturbance is 4 × 4.

The computer code for accomplishing this estimation process follows below.

```
N = 10;
c = 3*10^8;                   %speed of light

x1 = 7766188.44; y1 = -21960535.34;
z1 = 12522838.56;
X1 = [x1 y1 z1]';

x2 = -25922679.66; y2 = -6629461.28;
z2 = 31864.37;
X2 = [x2 y2 z2]';

x3 = -5743774.02; y3 = -25828319.92;
z3 = 1692757.72;
X3 = [x3 y3 z3]';

x4 = -2786005.69; y4 = -15900725.8;
z4 = 21302003.49;
X4 = [x4 y4 z4]';           % locations of visible satellites

d1 = 1022228206.42; d2 = 1024096139.11;
d3 = 1021729070.63; d4 = 1021259581.09;
d = [d1 d2 d3 d4]';          %pseudo distances, visible
                             satellites to receiver
```

```
x = 0; y = 0; z = 0; ctb = 0;        %initial guess for
receiver location and local clock error

X = [x y z]';                % receiver position vector
XX = [X' ctb]';               % state vector
A = eye(4);                   %propogation of X and ctb
P = 100000 * eye(4);          %cov of X and ctb
Q = 10000 * eye(4);           %process disturbance on X and ctb
R = 10 * eye(4);              % measurement noise in di's NxN

for j = 1:N

    xx(j) = [1 0 0] * X;
    yy(j) = [0 1 0] * X;
    zz(j) = [0 0 1] * X;
    ctbb(j) = ctb;                 %for plotting

  r1 = ((X - X1)' * (X - X1))^.5;
  r2 = ((X - X2)' * (X - X2))^.5;
  r3 = ((X - X3)' * (X - X3))^.5;
  r4 = ((X - X4)' * (X - X4))^.5;

h = [r1 - ctb r2 - ctb r3 - ctb r4 - ctb]';
    %h(x), di is the ith output

L(j) = (r1 - ctb - d1)^2 + (r2 - ctb -d2)^2 + (r3 - ctb - d3)^2 +
       (r4 - ctb - d4)^2;
        % indication of convergence of the process
H1 = [(1 / r1) * (X - X1)', - 1];
        %partial of h wrt x, N rows, 4 columns
H2 = [(1 / r2) * (X - X2)', - 1];
H3 = [(1 / r3) * (X - X3)', - 1];
H4 = [(1 / r4) * (X - X4)', - 1];
H = [H1; H2; H3; H4];
            %linearized output matrix
A = eye(4); %state transition matrix from linearization

P = A * P * A' + Q;
K = P * H' * inv(H * P * H' + R);
P = (eye(4) - K * H) * P;
residual = [d1 - h(1) d2 - h(2) d3 - h(3) d4 - h(4)]';
XX = [X; ctb];
```

```
XX = XX + K * (residual);
X = [eye(3); 0 0 0]' * XX;
ctb = [0 0 0 1] * XX;

end

figure
plot(L)

figure
plot(xx)
hold on
plot(yy)
hold on
plot(zz)
```

In Figure 5.8a and b are shown plots demonstrating the convergence of the estimates to the correct values. Note the values used for the initial covariance of state error and the covariances for the process disturbance and measurement noise.

Case 1

$P = 100,000 * eye(4)$, $Q = 10,000 * eye(4)$, $R = 10 * eye(4)$

$X = [-2,430,745.09594, \ -4,702,345.11359, \ 3,546,568.70600]'$

$tb = -3.33421$

Note the slight overshoot in each of the coordinate estimates.

Case 2

This example can be used to illustrate the impact of the values used for the covariances. To illustrate this point, the process is re-run with the initial covariance, P, reduced by a factor of 100, i.e., $P = 1,000 * eye(4)$, $Q = 10,000 * eye(4)$, $R = 10 * eye(4)$. In Figure 5.9a and b are shown plots demonstrating convergence of the estimates and the error function.

$X = [-2,430,745.09594, \ -4,702,345.11360, \ 3,546,568.70600]'$

$tb = -3.334215$

Note the absence of the overshoot in the coordinate estimates. Thus the transient portion of the result is quite different from Case 1, but the final values are the same.

Case 3

This example can be used to illustrate the importance of selecting appropriate values of R. Here it is increased by 1000: $P = 1,000 * eye(4)$, $Q = 100 * eye(4)$, $R = 10,000 * eye(4)$. In Figure 5.10a and b are shown plots of the results.

$X = [173,007,421.998, \ 391,056,363.126, \ -501,754,522.949]'$

(a)

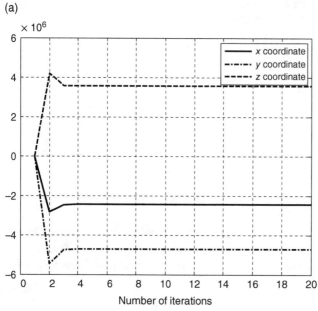

(b)

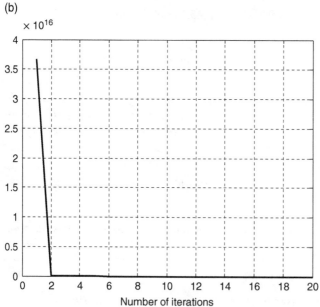

Figure 5.8 (a) Plot demonstrating the convergence of the coordinate estimates $P = 100,000 * eye(4)$, $Q = 10,000 * eye(4)$, and $R = 10 * eye(4)$. (b) Plot demonstrating the convergence of the error function $P = 100,000 * eye(4)$, $Q = 10,000 * eye(4)$, and $R = 10 * eye(4)$.

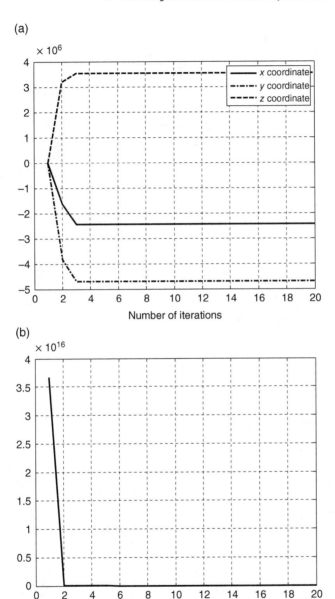

Figure 5.9 (a) Plot demonstrating the convergence of the coordinate estimates $P = 1{,}000 * \text{eye}(4)$, $Q = 10{,}000 * \text{eye}(4)$, and $R = 10 * \text{eye}(4)$. (b) Plot demonstrating the convergence of the error function $P = 1{,}000 * \text{eye}(4)$, $Q = 10{,}000 * \text{eye}(4)$, and $R = 10 * \text{eye}(4)$.

(a)

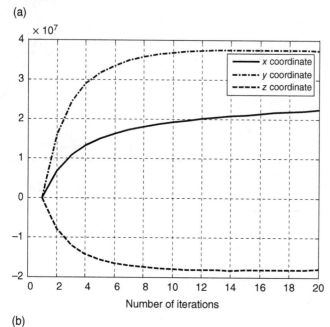

(b)

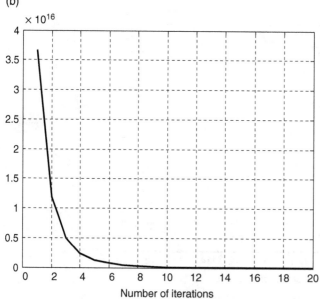

Figure 5.10 (a) Plot demonstrating lack of convergence of the coordinate estimates $P = 1,000 * \text{eye}(4)$, $Q = 100 * \text{eye}(4)$, and $R = 10,000 * \text{eye}(4)$. (b) Plot demonstrating lack of convergence of error function $P = 1,000 * \text{eye}(4)$, $Q = 100 * \text{eye}(4)$, and $R = 10,000 * \text{eye}(4)$.

The transient portion of the solutions is different from Case 1 or Case 2 and also after 10 iterations the solution has not converged to the proper solution.

These three executions of the Kalman Filter using different values for the various covariance matrix demonstrate how important these matrices are to the solutions. The proper adjustment of these covariance matrices is sometimes referred to as tuning.

General conclusions are that if Q is too small, there is slow convergence; however, large values for Q are okay. Since Q represents process disturbances, if it is made small, the estimate places its trust in the propagation equation, which here is the identity matrix and the state estimate tends to stay near its initial value longer.

Large values of R cause divergence, while setting R to zero is okay. The main source of information for solving the geolocation problem is the set of pseudo-distances to the visible satellites and the coordinates of these satellites. A large value of R corresponds to these pseudo-distance measurements and the satellite coordinates being unreliable, causing the estimation process to fail.

Regarding the covariance of the initial estimate, P, the final value of the state estimate is independent of the initial value of P, but using a value for P that is very small causes slower convergence. Starting with a small value for P is equivalent to telling the filter that the initial estimate of the state is a good one. This causes it to be slow to change even in the presence of measurements that yield large residuals. A large value for P works better.

For tracking a moving target, the initialization would provide a very good starting point for the next solution. Also, one would model the receiver with a different state equation to take into account its motion.

The Kalman Filter which was developed approximately half a century ago continues to play an important role in estimating the state of dynamic systems with limited and noisy measurements. Further examples of its application will appear in the chapters ahead.

Exercises

1 A A simple scalar dynamic system has the following discrete-time model:

$$x(k+1) = 0.5x(k) + u(k) + w(k)$$

with output

$$y(k) = x(k) + v(k)$$

Let the initial value of the state be

$$x(0) = 0$$

and let the input signal be

$$u(k) = 1; \; k = 0, 1, ..., 20$$

The standard deviation for both the process disturbance and the measurement noise is 0.1. Develop a simulation for this system and plot the state and the output versus k.

B Next develop a Kalman Filter for estimating the state of this system. Run the filter in parallel with the simulation. Plot the state, the output, and the estimate of state all versus k. Comment on the accuracy of the state estimate as compared with the output measurement.

C Repeat part (B), but now let the standard deviation for the measurement noise be 0.5. Again comment on the accuracy of the state estimate as compared with the output measurement.

2 A A simple scalar dynamic system with feedback has the following discrete-time model:

$$x(k+1) = 1.1x(k) + u(k) + w(k)$$

with output

$$y(k) = x(k) + v(k)$$

Let the initial value of the state be

$$x(0) = 3$$

The objective is to use feedback to stabilize the system utilizing a control signal given by

$$u(k) = -0.6y(k); \; k = 0, 1, ..., 20$$

The standard deviation for both the process disturbance and the measurement noise is 0.1.

Develop a simulation for this system and plot the state versus k.

B Next develop a Kalman Filter to operate in parallel with the system simulation and change the control algorithm to

$$u(k) = -0.6\hat{x}(k/k); \; k = 0, 1, ..., 20.$$

Compare the behavior here with that obtained in part A.

3 **A** A second-order dynamic system with feedback has the following discrete-time model:

$$\begin{bmatrix} x_1(k+1) \\ x_2(k+1) \end{bmatrix} = \begin{bmatrix} 1 & T \\ 0 & 1 \end{bmatrix} \begin{bmatrix} x_1(k) \\ x_2(k) \end{bmatrix} + \begin{bmatrix} T^2/2 \\ T \end{bmatrix} u(k) + \begin{bmatrix} T^2/2 \\ T \end{bmatrix} w(k)$$

with output

$$y(k) = \begin{bmatrix} 1 & 0 \end{bmatrix} x(k) + v(k)$$

Take T to be 0.1 second. The standard deviation for both the process disturbance and the measurement noise is 0.1. Let the initial value of the state be

$$\begin{bmatrix} x_1(0) \\ x_2(0) \end{bmatrix} = \begin{bmatrix} 0 \\ 0 \end{bmatrix}$$

The objective is to control the speed of the system using simple velocity feedback. One example of an idealized control signal would be given by

$$u(k) = 5(10 - x_2(k)) \quad k = 0,1,...,20$$

Since $x_2(k)$ is not measurable, this scheme is not feasible. Approximate $x_2(k)$ by the expression $[x_1(k) - x_1(k-1)]/T$. Develop a simulation for this system and plot the second component of state (velocity) versus k.

B Now develop a Kalman Filter to run in parallel with the system. Simulate and this time use the estimated value for velocity in the control algorithm, i.e., $u(k) = 10(5 - \hat{x}_2(k/k)) \quad k = 0,1,...,20$. Plot the second component of state versus k and compare the results with those of part A.

4 **A** Assume that for the front-wheel steered robot positions x and y and also heading are measurable. Show the behavior in going from the point 0,0 to the point 10,10. The initial heading angle is zero. There is no measurement noise and no process disturbance. Take as the desired heading $\psi_{des} = -\tan^{-1}\left(\frac{x_{des} - x}{y_{des} - y}\right)$ and use as the control algorithm

$$\alpha = Gain*(\psi_{des} - \psi) \text{ for } abs(Gain*(\psi_{des} - \psi)) \leq \pi/4$$

and

$$\alpha = \pi/4 \, sgn(\psi_{des} - \psi) \text{ for } abs(Gain*(\psi_{des} - \psi)) > \pi/4$$

Use the Euler method for simulating the system and take the sample interval to be 0.1 second. The length of the robot is 2 m. Take 0.5 as the Gain. Let the velocity be 1 m/s.

B Now assume there is no heading measurement. Use finite differencing of x and y at each sampling instant to approximate heading, i.e.,

$$\psi \approx -\tan^{-1}\left(\frac{x(k)-x(k-1)}{y(k)-y(k-1)}\right)$$

and control the robot with the same algorithm as in the first part. Assume perfect measurements on x and y and assume zero process disturbance.

C Repeat the second part except now include a random process disturbance in the steering angle for the simulation and include random noise in the x and y measurements. See what type of performance the finite differencing method for heading determination provides in the presence of noise. Let the standard deviation of the disturbance be 0.05 rad and let the standard deviation of the measurement noise be 0.2 m for x and y.

D Now re-solve the problem of control with no heading measurement, this time by implementing a Kalman Filter. Your simulation should include the process disturbance and the measurement noise as in the third part. Since the process disturbance is not measurable directly, do not include it in the prediction equation of the filter. Compare the behavior obtained in parts three and four, and see if either approaches the ideal behavior obtained in part one.

Discuss your results.

5 A plant model for a steel-bending operation is given by

$$\begin{bmatrix} x_1 \\ x_2 \\ x_3 \end{bmatrix} = \begin{bmatrix} 0 & 1 & 0 \\ -1.8 & -0.7 & 1.1 \\ 0 & 0 & 0 \end{bmatrix} \begin{bmatrix} x_1 \\ x_2 \\ x_3 \end{bmatrix} + \begin{bmatrix} 0 \\ 0 \\ 0.9 \end{bmatrix} u + \begin{bmatrix} 0 \\ 0 \\ 0 \end{bmatrix} w \text{ with output}$$

$$\begin{bmatrix} y_1 \\ y_2 \end{bmatrix} = \begin{bmatrix} 1 & 0 & 0 \\ 0 & 0 & 1 \end{bmatrix} \begin{bmatrix} x_1 \\ x_2 \\ x_3 \end{bmatrix} + \begin{bmatrix} v_1 \\ v_2 \end{bmatrix}$$

Here the x_1 represents curvature of the bent steel, x_2 represents curvature rate, and x_3 represents the transverse roller position. Develop a discrete-time state model with a sampling interval of 0.1 second. Next develop a Kalman Filter for estimating all the states. This is an example where the Kalman Filter can be used to estimate the components of state that are not directly measurable and also to give improved estimates of those that are measurable. Let the covariances for w, v_1, and v_2 all be taken as 0.1. Let the input signal be a unit step. Simulate the system along with the Kalman Filter and plot the actual states, the measurements of states 1 and 3, and all three state estimates.

6 For the exercise described above, repeat everything, except now implement the control according to $u = 26.9 \times$ reference curvature $- 28.5x_1 - 23x_2 - 9.5x_3$. Let the reference curvature be a step function.

References

Brogan, W. L., *Modern Control Theory*, Prentice Hall, Upper Saddle River, NJ, 1991.

Brown, Robert Grover and Hwang, Patrick Y. C., *Introduction to Random Signals and Applied Kalman Filtering*, 3rd Edition, Wiley, New York, NY, 1996.

Bryson, A. E. and Ho, Y.-C., *Applied Optimal Control: Optimization, Estimation and Control*, Blaisdell, Waltham, MA, 1969.

Cook, G. and Dawson, D. E., "Optimum Constant-Gain Filters," *IEEE Transactions on Industrial Electronics and Control Instrumentation*, Vol. **21**, No. 3 (August 1974), pp. 159–163.

Cook, G. and Lee, Y. C., "Use of Kalman Filtering for Data Resolution Enhancement," *IEEE Transactions on Industrial Electronics and Control Instrumentation*," Vol. **22**, No. 4 (November 1975), pp. 497–500.

Grewal, Mohinder S. and Angus, P., *Andrews "Kalman Filtering,"* 3rd Edition, Wiley, 2008.

Han, S., Zhang, Q., and Noh, H., "Kalman Filtering of DGPS Positions for a Parallel Tracking Application," *Transactions of the ASAE*, Vol. **45**, No. 3 (2002), pp. 553–559.

Haykin, S., *Adaptive Filter Theory*, Prentice Hall, 2002.

Julier, S. J. and Uhlmann, J. K., "A New Extension of the Kalman Filter to Nonlinear Systems," *The 11th International Symposium on Aerospace/Defense Sensing, Simulation and Controls, Multi Sensor Fusion, Tracking and Resource Management, SPIE*, Vol. 3 (pp. 182–193, Orlando, FL, April 20–25, 1997).

Kiriy, E. and Buehler, M., *Three-State Extended Kalman Filter for Mobile Robot Localization*, Tech. Rep., McGill University, Montreal, 2002.

Meditch, J. S., *Stochastic Optimal Linear Estimation and Control*, McGraw-Hill, New York, NY, 1969.

Roumeliotis, S. and Bekey, G., "Bayesian Estimation and Kalman Filtering: A Unified Framework for Mobile Robot Localization," *Proceedings of IEEE International Conference on Robotics and Automation* (pp. 2985–2992, San Francisco, CA, April 24–28, 2000).

Sage, A. P. and Melsa, J. L., *Estimation Theory with Applications to Communication and Control*, McGraw-Hill Inc, 1971.

Trostmann, E., Hansen, N. E., and Cook, G., "General Scheme for Automatic Control of Continuous Bending of Beams," *Transactions of the ASME, Journal of Dynamic Systems, Measurement and Control*, Vol. **104**, No. 2 (June 1982), pp. 173–179.

6

Remote Sensing

6.1 Introduction

This chapter is devoted to the acquisition of images via sensors mounted on a mobile robot. Particular attention is given to projecting the sensor field of view onto a surface (such as the ground) and then converting a specific pixel coordinate (for example the coordinates of a detected object of interest) of this field into the actual ground coordinates. Pointing at a detected target for ranging is also treated.

6.2 Camera-Type Sensors

In this section, we analyze camera-type sensors and address the task of converting sensor frame coordinates of objects to their representation in other frames such as vehicle coordinates and earth coordinates. Figure 6.1 illustrates these multiple frames. The analysis applies to infrared cameras and also to visible wavelength cameras. The only requirement is that it be a pinhole-type digital camera with the field of view divided into pixels. In order to proceed, we first define some coordinate frames. The sensor coordinate frame has y aligned with the longitudinal axis of the camera, i.e., the camera boresight, x pointing to the right, and z pointing upward. We seek to compute the direction of the vector pointing at the target, which appears in the ith row and jth column of the pixel array. See Figure 6.2.

By expressing the pixel coordinates in the virtual focal plane rather than in the actual focal plane, one eliminates the need to think in terms of reversed coordinate directions.

The coordinates shown in Figure 6.3 are the coordinates of the center of the pixel in question. The rows, i, and the columns, j, are numbered in the same way as rows and columns of a matrix, i.e., beginning at the upper left corner. We can

Mobile Robots: Navigation, Control and Sensing, Surface Robots and AUVs,
Second Edition. Gerald Cook and Feitian Zhang.
© 2020 by The Institute of Electrical and Electronics Engineers, Inc.
Published 2020 by John Wiley & Sons, Inc.

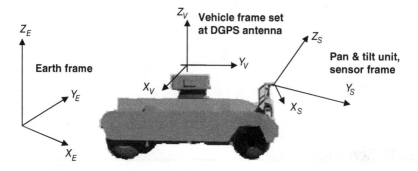

Figure 6.1 Sensor-bearing mobile robot.

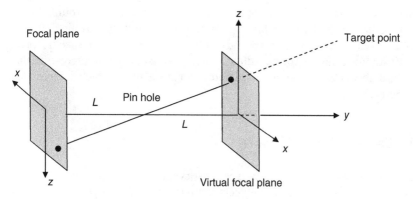

Figure 6.2 Diagram of a pinhole camera.

compute y and z as follows. Assume that the pixel array has N rows and M columns. The pixel size is Δw wide by Δh high. The focal length is L. The dimensions of the field of view are given by *VFOV* and *HFOV*. Here *VFOV*/2 and *HFOV*/2 are each half of the respective sensor's vertical field of view and horizontal field of view, respectively. By setting $j = 1$ and using trigonometric arguments we have

$$\frac{[(M-1)/2]\Delta w}{L} = \tan{(HFOV/2)}$$

or

$$\frac{1}{L} = \frac{\tan{(HFOV/2)}}{[(M-1)/2]\Delta w}$$

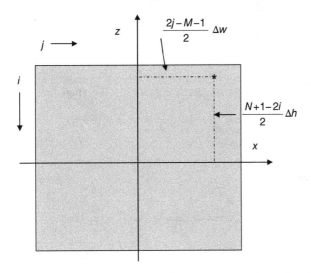

Figure 6.3 Computing *X–Z* coordinates for *ij*th pixel in a digital camera using the virtual focal plane.

Now noting that one may express $x = \dfrac{2j - M - 1}{2}\Delta w$ and $y = L$ we have

$$x/y = x/L = \frac{[(2j - M - 1)/2]\Delta w}{L}$$

or

$$x/L = \frac{2j - M - 1}{M - 1}\tan\left(HFOV/2\right) = x_{NRO} \tag{6.1}$$

where x_{NRO} is defined as the normalized horizontal readout. Likewise in the z coordinate setting $j = 1$ yields

$$\frac{[(N - 1)/2]\Delta h}{L} = \tan\left(VFOV/2\right)$$

or

$$\frac{1}{L} = \frac{\tan\left(VFOV/2\right)}{[(N - 1)/2]\Delta h}$$

Here noting that $z = \dfrac{N + 1 - 2i}{2}\Delta h$ and again using $y = L$

$$z/y = z/L = \frac{[(N + 1 - 2i)/2]\Delta h}{L}$$

or

$$z/L = \frac{N+1-2i}{N-1} \tan\left(VFOV/2\right) = z_{NRO} \tag{6.2}$$

where z_{NRO} is defined as the normalized vertical readout. We then have as azimuth

$$\psi_{ray} = \tan^{-1}\frac{-x}{L} = -\tan^{-1}x_{NRO} \tag{6.3}$$

and for elevation we have

$$\theta_{ray} = \tan^{-1}\frac{z}{\sqrt{L^2+x^2}} = \tan^{-1}\frac{z/L}{\sqrt{1+x^2/L^2}} = \tan^{-1}\frac{z_{NRO}}{\sqrt{1+x_{NRO}^2}} \tag{6.4}$$

The steps are to first determine the pixel coordinates of the object of interest. Then x_{NRO} is calculated from equation (6.1) and z_{NRO} is calculated from equation (6.2). Equations (6.3) and (6.4) then yield azimuth and elevation of the ray in camera coordinates.

Two samples of IR images are shown in Figure 6.4. The vertical field of view of the camera was approximately 12° and the horizontal field of view was approximately 16°. The corresponding pixel dimensions were 240 by 320. The image on the left is a section of earth that has been recently disturbed and is still loose. The image on the right is a road with markers placed along the right side. Through the use of a software tool, one can click on an object of interest within an image such as these and obtain its pixel coordinates, i and j. From these, one can use the equations just developed and determine the azimuth and elevation (i.e., the direction in camera coordinates) of the ray pointing at this object.

Figure 6.4 Sample IR images.

It is now desired to express the ground location of the tip of this ray that passes through the *ij*th pixel in sensor coordinates, i.e., the point where this ray intersects the ground. The coordinates are given by

$$
\begin{bmatrix} x \\ y \\ z \end{bmatrix}_{snsor\ coords} = \begin{bmatrix} -r_{ij} \cos \theta_{ray} \sin \psi_{ray} \\ r_{ij} \cos \theta_{ray} \cos \psi_{ray} \\ r_{ij} \sin \theta_{ray} \end{bmatrix}
\tag{6.5}
$$

It is seen that the range to the pixel in question (r_{ij}) is required for this calculation. If the terrain in front of the vehicle is planar and this plane extends underneath the vehicle, and the boresight of the camera is parallel to the earth (zero pitch for the camera), then the range r_{ij} may be calculated by solving the equation

$$r_{ij} \sin \theta_{ray} = -H$$

or

$$
r_{ij} = -H / \sin \theta_{ray}
\tag{6.6}
$$

where H is the height of the camera above the ground. Knowing r_{ij}, one can now compute the x and y coordinates of the object in camera coordinates. These can then be converted to vehicle coordinates and finally to earth coordinates.

The above is the simplest case possible and is presented to introduce the idea of computing the intersection of the ray with the ground as a means of determining its length. The more complex case of arbitrary vehicle orientation is now addressed. Here the vehicle may have nonzero pitch and roll as well as yaw, and the camera may be mounted with arbitrary orientation as well. In this case we must use a rotation matrix and express the ray in vehicle axes and finally earth axes before solving for r.

The vector expressed in camera coordinates is given as

$$
\begin{bmatrix} X \\ Y \\ Z \end{bmatrix}_{ray\ cam\ coords} = \begin{bmatrix} -r \cos \theta \sin \psi \\ r \cos \theta \cos \psi \\ r \sin \theta \end{bmatrix}_{ray\ cam\ coords} = r \begin{bmatrix} \hat{X} \\ \hat{Y} \\ \hat{Z} \end{bmatrix}_{ray\ in\ cam\ coords}
\tag{6.7a}
$$

where r is yet to be determined. In vehicle coordinates this becomes

$$
\begin{bmatrix} X \\ Y \\ Z \end{bmatrix}_{ray\ in\ veh\ coords} = rR_{cam-veh} \begin{bmatrix} \hat{X} \\ \hat{Y} \\ \hat{Z} \end{bmatrix}_{ray\ in\ cam\ coords} + \begin{bmatrix} X \\ Y \\ Z \end{bmatrix}_{cam\ origin\ veh\ coords}
\tag{6.7b}
$$

Finally converting this to earth coordinates we have

$$
\begin{bmatrix} X \\ Y \\ Z \end{bmatrix}_{ray\ earth\ coords} = r R_{veh-earth} R_{cam-veh} \begin{bmatrix} \hat{X} \\ \hat{Y} \\ \hat{Z} \end{bmatrix}_{ray\ cam\ coords}
$$

$$
+ R_{veh-earth} \begin{bmatrix} X \\ Y \\ Z \end{bmatrix}_{cam\ origin\ veh\ coords} + \begin{bmatrix} X \\ Y \\ Z \end{bmatrix}_{veh\ origin,\ earth\ coords} \tag{6.7c}
$$

Setting the z-component of the ray in earth coordinates equal to the z-component of the vehicle origin in earth coordinates minus the height of the vehicle origin above the ground expressed in earth coordinates, one can solve for r, the length of the vector. (Note that the placement of the GPS antenna on the vehicle is defined as the vehicle origin. Thus, the height of the vehicle origin is the same as the height of the GPS antenna.)

$$
r[0\ 0\ 1] R_{veh-earth} R_{cam-veh} \begin{bmatrix} \hat{X} \\ \hat{Y} \\ \hat{Z} \end{bmatrix}_{ray\ can\ coords}
$$

$$
+ [0\ \ 0\ \ 1] R_{veh-earth} \begin{bmatrix} X \\ Y \\ Z \end{bmatrix}_{cam\ origin,\ veh\ coords} + [0\ \ 0\ \ 1] \begin{bmatrix} X \\ Y \\ Z \end{bmatrix}_{veh\ origin,\ earth\ coords}
$$

$$
= [0\ \ 0\ \ 1] \begin{bmatrix} X \\ Y \\ Z \end{bmatrix}_{veh\ origin,\ earth\ coords} - [0\ \ 0\ \ 1] R_{veh-earth} \begin{bmatrix} 0 \\ 0 \\ H_{ant} \end{bmatrix} \tag{6.8a}
$$

This simplifies to

$$
r[0\ \ 0\ \ 1] R_{veh-earth} R_{cam-veh} \begin{bmatrix} \hat{X} \\ \hat{Y} \\ \hat{Z} \end{bmatrix}_{ray\ can\ coords}
$$

$$
= -[0\ \ 0\ \ 1] R_{veh-earth} \left(\begin{bmatrix} X \\ Y \\ Z \end{bmatrix}_{cam\ origin,\ veh\ coords} + \begin{bmatrix} 0 \\ 0 \\ H_{ant} \end{bmatrix} \right) \tag{6.8b}
$$

which is a scalar equation in r that one can solve by division. Using this value for r, one can then express the location of an object of interest in any of the coordinate systems using the equations above. This all assumes that the elevation of the earth at the point where the ray intersects the earth is the same as that under the vehicle or robot.

Example 1   *Consider a camera with pixel dimensions $N = 360$ and $M = 480$. Take as the field of view, HFOV = $\pi/4$ and VFOV = $\pi/6$. The camera's longitudinal axis or boresight is at a pitch angle of $\theta = -\pi/8$ and the height of the camera is 2 m. Determine the footprint, i.e., the intersection of the field of view with the ground.*

Solution 1

```
% Every 30th pixel is plotted via the computer program
with code given below.
 N = 480; M = 360;
 HFOV = pi / 4; VFOV = pi / 6;
 H = 2;            % height of camera

thetac = -pi / 8; %camera boresight pitch angle with
respect to robot
% camera yaw and roll wrt robot assumed to be zero.
sir = 0;          % yaw of robot wrt earth
thetar = 0;       % pitch of robot wrt earth
phir = 0;         % roll of robot wrt earth

%camera position wrt robot coordinate center which
coincided with gps antenna
xc = 0; yc = 0.5; zc = -0.25;
Xc = [xc yc zc]';

%robot position in earth coordinates, for simplicity
taken to be zero here
xe = 0; ye = 0; ze = 0;
Xr = [xe ye ze]';

%rotation camera wrt robot
Royc = [1 0 0; 0 1 0; 0 0 1];
Ropc = [1 0 0; 0 cos(thetac) -sin(thetac); 0 sin(thetac)
cos(thetac)];
```

```
Rorc = [1 0 0; 0 1 0; 0 0 1];
Roc = Royc * Ropc * Rorc;

%rotation robot wrt earth
Royv = [cos(sir) -sin(sir) 0; sin(sir) cos(sir) 0; 0 0 1];
Ropv = [1 0 0; 0 cos(thetar) -sin(thetar); 0 sin(thetar)
cos(thetar)];
Rorv = [ cos(phir) 0 sin(phir); 0 1 0; - sin(phir) 0
cos(phir)];
Rov = Royv * Ropv * Rorv;
figure
for ii = 1:16
    for jj = 1:12
        i = ii * 30; j = jj * 30;
        xn = ((2 * j - M - 1) / (M - 1)) * tan(HFOV / 2);
        zn = ((N + 1 - 2 * i) / (N - 1)) * tan(VFOV / 2);
        si = -atan(xn);              %azimuth of ray
corresponding to pixel
            theta = atan(zn / sqrt(1 + yn^2));        %
elevation of ray corresponding to pixel
        xhat = -cos(theta) * sin(si);      %unit vector in
sensor frame
        yhat = cos(theta) * cos(si);
        zhat = sin(theta);
        Xhat = [xhat yhat zhat]';
        num = -[0 0 1] * Rov * (Xc + [0 0 H]');      %range
computation
        denom = [0 0 1] * Rov * Roc * Xhat;
        range = num / denom;
        Xte = range * Rov * Roc * Xhat + Rov * Xc + Xr; %
target in earth coords
        xx = Xte(1);
        yy = Xte(2);
        plot(xx, yy, '*')              % plots ground
coordinates for each pixel computed
        hold on
    end
end
```

The projection of the set of pixels onto the earth is shown in Figure 6.5.

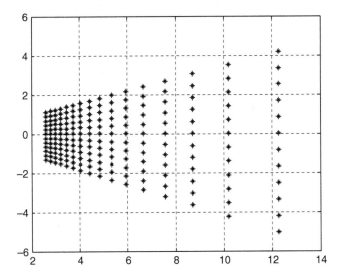

Figure 6.5 Camera footprint with vehicle at zero yaw, pitch, and roll.

Example 2 *If the vehicle is tilted, this will be reflected in the vehicle to earth rotation. Repeat the previous example, but now suppose the vehicle itself is pitched downward by an amount $\pi/20$ radians or $9°$. Determine the footprint for this case.*

Solution 2

The field of view is computed using the code of the previous example. Figure 6.6 shows that the farthest extremity of the field of view has shortened from 11.5 to 5.5 m as a result of the vehicle being pitched downward.

Example 3 *If the vehicle is rolled, then the transformation performed on the ray before computing range must also include the roll. Consider the previous example, but now with the vehicle rolled positively (left side up) by an amount $\pi/18$ radians or $10°$.*

Solution 3

The camera footprint for the case where the vehicle is rolled in the positive direction by $\pi/18$ radians is shown in Figure 6.7.

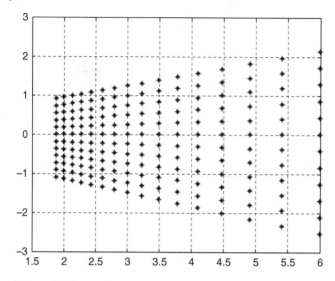

Figure 6.6 Camera footprint with vehicle at zero yaw and roll, but pitched down $\pi/20$ radians.

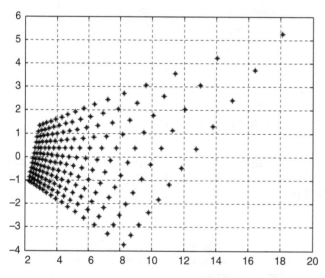

Figure 6.7 Camera footprint with vehicle at zero yaw and pitch, but rolled by $\pi/18$ radians.

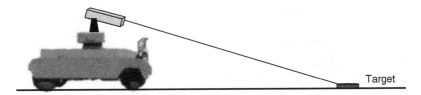

Figure 6.8 Illustration of single-camera algorithm for determining length of ray to target.

By observing how the fields of view for the camera-type sensors change so drastically with vehicle attitude, it is apparent that accurate instrumentation for this purpose is absolutely necessary. Yaw, pitch, and roll, all have a major impact on the accurate geolocation. Possible instrumentation includes an array of GPS antennas or an inertial measurement unit. An inclinometer could be used subject to the limitation that there be no vehicle acceleration during sensing. Terrain elevation also impacts geolocation accuracy. Figure 6.8 illustrates the geometry for determining major errors.

Since the intersection of ray with the ground is computed based on the angle of the ray, the height of the antenna, and the elevation of the earth, this intersection depends on vehicle attitude (measured), vehicle position (measured), and terrain elevation (either assumed flat or obtained from range scan).

The sensitivity functions when using the algorithm based on the intersection of the ray with the ground are

$$\sigma^2_{dnrng} = \left(\frac{r^2}{H_{antenna}}\right)^2 \sigma^2_{pitch} + \left(\frac{r}{H_{antenna}}\right)^2 \sigma^2_{elevation} \qquad (6.9a)$$

$$\sigma^2_{crsrng} = r^2 \sigma^2_{heading} \qquad (6.9b)$$

While the coefficient of σ^2_{pitch} is typically larger than the coefficient of $\sigma^2_{elevation}$, the error in elevation itself is much larger than the error in pitch causing both terms to contribute significantly. Thus, very precise position and orientation information is required for accurate determination of the location of a detected object.

6.3 Stereo Vision

Because of the potential errors when using a single camera, stereo vision is also considered as a means of ground registration of objects of interest. Here we use a pair of cameras in a coordinated way. The diagram in Figure 6.9 illustrates the setup. Both cameras are mounted rigidly on a single platform and spaced apart

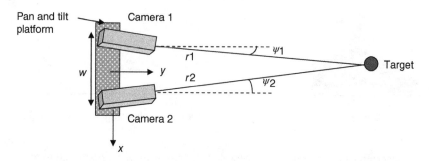

Figure 6.9 Setup for using stereo vision for geolocating objects of interest.

by a known quantity. This platform could be attached to a pan and tilt unit as a means of scanning with the cameras.

Once an object of interest is identified, one determines the angle of the respective rays from each camera to the object. These include angles of the rays in camera coordinates and also the fixed angles of the cameras with respect to the coordinate frame of the mounting platform. Both azimuth and elevation are needed. Next, one uses the law of sines to determine the distances from each camera to the point of intersection of the rays. The equations are

$$r_1 = W \frac{\cos \psi_2}{\sin (\psi_1 + \psi_2) \cos \theta_1} \tag{6.10a}$$

and

$$r_2 = W \frac{\cos \psi_1}{\sin (\psi_1 + \psi_2) \cos \theta_2} \tag{6.10b}$$

With this information, one can compute the coordinates of the object in the frame of the camera mounting platform. In terms of r_1 these are

$$x = -\frac{W}{2} + r_1 \sin \psi_1 \cos \theta_1 \tag{6.11a}$$

$$y = r_1 \cos \psi_1 \cos \theta_1 \tag{6.11b}$$

and

$$z = r_1 \sin \theta_1 \tag{6.11c}$$

In terms of r_2 these become

$$x = \frac{W}{2} - r_2 \sin \psi_2 \cos \theta_2 \tag{6.12a}$$

$$y = r_2 \cos \psi_2 \cos \theta_2 \qquad (6.12b)$$

and

$$z = r_2 \sin \theta_2 \qquad (6.12c)$$

Clearly, both solutions should place the object at the same point. The error variances when using stereo vision are

$$\sigma^2_{downrange} = \left(\frac{r^2}{W}\right)^2 \left(\sigma^2_{\psi_1} + \sigma^2_{\psi_2}\right) + \left(\frac{r}{W}\right)^2 \sigma^2_W \qquad (6.13a)$$

and

$$\sigma^2_{crossrange} = r^2 \sigma^2_{heading} \qquad (6.13b)$$

Since W is precisely known, this reduces to

$$\sigma^2_{downrange} = \left(\frac{r^2}{W}\right)^2 \left(\sigma^2_{\psi_1} + \sigma^2_{\psi_2}\right) \qquad (6.14a)$$

and

$$\sigma^2_{crossrange} = r^2 \sigma^2_{heading} \qquad (6.14b)$$

Here σ_ψ is comprised primarily of the uncertainty in camera mounting angles and is expected to be extremely small. Further, the effect of terrain is eliminated, significantly reducing the downrange errors.

Example 4 *Two cameras for stereo imaging are mounted on a platform 2 m apart. An object is detected in each camera and determined to be the same object. For camera 1, based on the pixel coordinates and the camera mounting angles, the azimuth angle of the ray to the target is determined to be 0.3 rad to the right. The pitch angle of the ray is −0.2383 rad. For camera 2, based on the pixel coordinates and the camera mounting angles, the azimuth angle for the ray to the target is 0.2 rad to the left and the pitch angle is −0.2466. Determine the location of the object with respect to the camera mounting platform.*

Solution 4

Using the formula for r_1 given above,

$$r_1 = W \frac{\cos \psi_2}{\sin (\psi_1 + \psi_2) \cos \theta_1}$$

yields

$$r_1 = 2\frac{\cos{(0.2)}}{\sin{(0.3 + 0.2)}\cos{(-0.2383)}} = 4.21$$

Thus, in the camera mounting platform coordinates we obtain

$$x = -\frac{W}{2} + r_1 \sin{\psi_1}\cos{\theta_1} = 0.21$$

$$y = r_1 \cos{\psi_1}\cos{\theta_1} = 3.91$$

and

$$z = r_1 \sin{\theta_1} = -1$$

Similarly, solving for r_2 via the formula for it

$$r_2 = W\frac{\cos{\psi_1}}{\sin{(\psi_1 + \psi_2)}\cos{\theta_2}}$$

yields

$$r_2 = 2\frac{\cos{(0.3)}}{\sin{(0.5)}\cos{(-0.2466)}} = 4.11$$

Now using the expressions involving r_2 gives

$$x = \frac{W}{2} - r_2 \sin{\psi_2}\cos{\theta_2} = 0.21$$

$$y = r_2 \cos{\psi_2}\cos{\theta_2} = 3.91$$

and

$$z = r_2 \sin{\theta_2} = -1$$

which are in agreement with the other results. This location may now be transformed to the robot coordinates and then to earth coordinates.

One important consideration when using stereo vision is making sure that one has detected the same target in both images. Whatever image recognition method is used for the detection in one frame would be used in the other, and the same features would be sought in each frame. However, there is another factor which greatly alleviates the problem of locating the target in the second frame. If the two cameras are mounted on the mounting frame at the same elevation angle and if their longitudinal axes are almost parallel, then the target will appear at almost the same pixel row in each camera. Thus, from the pixel coordinates of the detection in the first image frame, one has a very good starting point for the search in the second image frame.

6.4 Radar Sensing: Synthetic Aperture Radar

Radar sensing relies on the transmission of electromagnetic energy and the measurement of the returned signal. The synthetic aperture method of sensing utilizes multiple signal returns from each pixel to be imaged. These multiple returns may be obtained from a single antenna, which moves along and is in a different location for each return, or they may be obtained from an array of sensors with each sensor at a different location. The actual radar signal may be a pulse or sinusoids of stepped frequency. If a stepped frequency signal is employed, the received signal is converted to the time domain by use of the inverse Fourier Transform. For a moving receiver antenna, this conversion is done for each antenna location. If an array of receiver antennas is used, this is done for each antenna. Here we consider a horizontal array of M receiver antennas mounted on the vehicle. At the mth receiver, the received signal is converted to the time domain via the inverse discrete Fourier Transform

$$g_m(t) = \frac{1}{N_f} \sum_{p=1}^{N_f} G_m(f_p) \exp(i2\pi f_p t) \tag{6.15}$$

A possible graph of such a signal is shown in Figure 6.10.

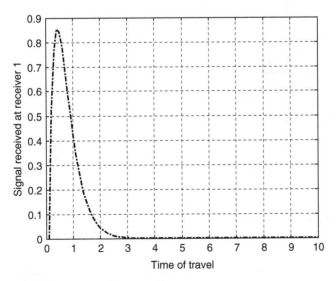

Figure 6.10 Plot of hypothetical signal received at receiver 1 versus time of travel.

Now to compute the intensity of signals returned from the pixel (x_F, y_F), one first computes the travel time from the single transmitter, to the pixel in question and back to each receiver element. For the mth receiver element this is

$$
t_{ground}^m (x_F, y_F) = \frac{1}{c} \left[(x_{s,m} - x_F)^2 + (y_{s,m} - y_F)^2 + z_{s,m}^2 \right]^{1/2}
$$
$$
+ \frac{1}{c} \left[(x_{r,m} - x_F)^2 + (y_{r,m} - y_F)^2 + (z_{r,m} - z_F)^2 \right]^{1/2}
\tag{6.16}
$$

That is, the distance is computed from the transmitter to the pixel of interest to each receiver. This distance is converted to travel time by dividing by the speed of light.

The time-domain waveform received at each receiver element is then evaluated at the time computed for the pixel of interest and for the respective receiver element. Hypothetical samples of waveforms at the different receiver antennas and travel times associated with a particular pixel are illustrated in Figures 6.11–6.13. These signal values are then summed over all receiver elements

$$
S_n(x_F, y_F) = \frac{1}{M} \sum_{m=1}^{M} g_m \left[t_{system} + t_{ground}^m (x_F, y_F) \right]
\tag{6.17}
$$

The result is the signal intensity corresponding to the pixel being imaged. Clearly, there is other signal content included in each of the terms of the summation above. Each term includes all reflection from an ellipse passing through

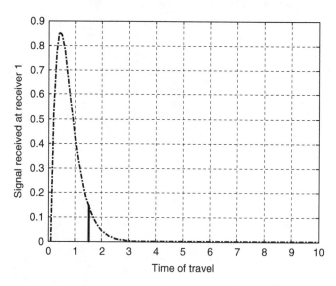

Figure 6.11 Plot of the hypothetical signal received at receiver 1, travel time for the pixel of interest noted.

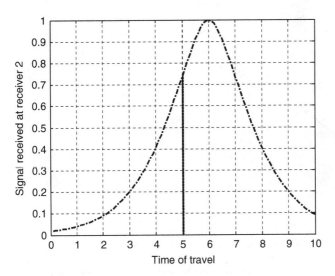

Figure 6.12 Plot of the hypothetical signal received at receiver 2, travel time for the pixel of interest noted.

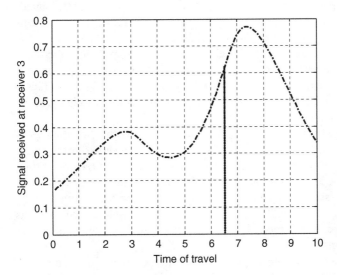

Figure 6.13 Plot of hypothetical signal received at receiver 3, travel time for the pixel of interest noted.

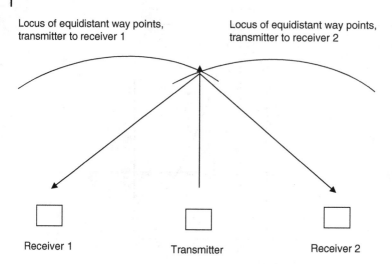

Locus of equidistant way points, transmitter to receiver 1

Locus of equidistant way points, transmitter to receiver 2

Receiver 1

Transmitter

Receiver 2

Figure 6.14 Illustration of sources of received signals using time gating. Only the point where the ellipses intersect is common to both receivers. Signals reaching the receivers via other points on the ellipses are incoherent.

the pixel of interest. These ellipses, one corresponding to each receiver element, all coincide at the pixel of interest. Therefore, these components of the signals add coherently whereas the other components are incoherent and tend to cancel each other. This point is illustrated in Figure 6.14.

As an aside, one can see that the angle at which the ellipses intersect each other depends on the width of the array. The narrower the array, the more difficult it is to determine the intersection point in the cross-track direction. This is the reason that the finite width of the antenna array limits the resolution in the cross-track coordinate.

This process is continued for each pixel in the frame. Note that first the signals are transmitted by the transmitter and collected by all receivers over the entire frequency range. The inverse DFT of the signal received at each receiver is then computed. Then the images are formed afterward for the entire frame, pixel-by-pixel. Parallel processing may be used here. Signals for the next frame may be transmitted and collected while the image is being formed for the previous frame. The required time interval between frames would be the maximum of these two operations.

Because of the way the synthetic aperture radar (SAR) images are formed, the image initially obtained is in vehicle coordinates. If the vehicle is yawed, pitched, or rolled with respect to the field of view to be imaged, this needs to be taken into account for exact computation of the distances from the transmitting antenna to the pixel in question and back to each receiver element. Likewise, if the terrain in front of the vehicle is uneven, this also needs to be incorporated for exact distance calculations. Fortunately for radar, the effects of vehicle pitch and roll are minimal in the range calculation as are the effects of uneven terrain. The two most important pieces of information are vehicle position and heading or yaw.

Effects of uneven terrain on ground registration may be analyzed in more detail utilizing Figure 6.15. With the receiver elements arranged in a horizontal line with the transmitter centered on this array, the locus of points lying at the intersection of the ellipses is as shown below. Here the receiver elements are located along the x axis of the antenna and symmetric about the z axis. The transmitter is located at the origin of these axes. The ray to the pixel of interest is labeled with length R. Note that this ray makes angle alpha with respect to the yz plane. The equations for signal time-of-travel are such that one can rotate this ray about the x axis, maintaining the angle alpha with respect to the yz plane, and define a circle such that reflections from any point lying on this circle will be imaged. If the height of the antenna above the terrain being imaged is H, as was expected, then the pixel's y and z coordinates will be as was expected. However, if the terrain is higher or lower than was expected, the y and z coordinates of the pixel imaged will change. The x coordinate is unchanged by elevation changes. Knowledge of the terrain elevation can be used to prevent this type of error in image formation. Fortunately the y coordinate changes only very slightly with elevation causing errors due to unknown terrain to be minimal.

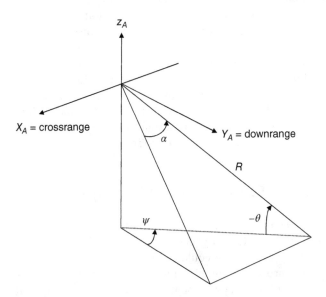

Figure 6.15 Geometrical considerations, radar-type sensor.

The equations for the point imaged are as follows.

$$x_A = -R \sin \alpha \tag{6.18a}$$

$$z_A = -H \tag{6.18b}$$

and

$$y_A = \sqrt{R^2 \cos^2 \alpha - H^2} \tag{6.18c}$$

For geolocation of targets, the orientation of the field of view with respect to the vehicle must be known. Typically for forward-looking sensing it would be prespecified as a rectangle in front of the vehicle. The actual design of the radar would be dictated by this specification. Thus each pixel has a known pair of x and y coordinates with respect to vehicle coordinates. The z coordinate will be obtained either by assumption of a flat earth or from a terrain model based on an array of range measurements. To convert from pixel location in vehicle coordinates to pixel location in earth coordinates, one uses the equation below

$$\begin{bmatrix} X_{PE} \\ Y_{PE} \\ Z_{PE} \end{bmatrix} = \begin{bmatrix} X_{VE} \\ Y_{VE} \\ Z \end{bmatrix} + R_{VE} \begin{bmatrix} X_{PV} \\ Y_{PV} \\ Z_{PV} \end{bmatrix} \tag{6.19}$$

For the rotation matrix (R_{VE}), one needs to know the vehicle attitude. Of the three components (yaw, pitch, and roll), yaw is the most crucial for accurate geolocation when using radar. As stated earlier, the geolocation accuracy for radar is fairly insensitive to errors in pitch and roll. This is in contrast to camera-type sensors, which are quite sensitive to all three components of attitude. Candidate sensors for attitude measurement include a digital compass, an array of GPS antennas, or an inertial measurement unit. An inclinometer could be a candidate if there is no acceleration during the imaging, e.g., imaging is done with the vehicle stopped. Clearly, one also needs vehicle position, which would most likely be obtained via DGPS or from an inertial navigation system. See Kositsky and Milanfar for an example of this type of radar.

6.5 Pointing of Range Sensor at Detected Object

If an object of interest has been detected with only an IR camera, one may know quite precisely the direction to the target but not the distance (range). A combination of an IR camera and radar could be used for improving the precision of the georegistration process. A combination of the IR camera and a ranging device such as a ranging laser may also be used to complete the required measurements. Assuming that an object has been detected via a camera and the

Figure 6.16 P&T unit at zero yaw and zero pitch, object of interest detected off the camera boresight.

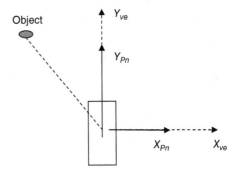

direction to it has been determined, we now turn to the task of pointing a ranging laser in that direction. See Figure 6.16.

We assume that the ranging laser and the camera are both mounted on a pan and tilt device and that they are coaligned with each other. We further assume at this point that the pan and tilt angles are both zero at the time of target detection, i.e., the camera is aimed straight ahead. Later we will handle the more general case.

From Section 6.1 we may compute the direction of the ray to the target corresponding to the location of the target in the image, i.e., corresponding to the *ij*th camera pixel coordinate. Using these angles computed as has been illustrated, one gets as an expression for the unit ray pointing at the target

$$x = -\sin \psi_{ray} \cos \theta_{ray} \tag{6.20}$$

$$y = \cos \psi_{ray} \cos \theta_{ray} \tag{6.21}$$

and

$$z = \sin \theta_{ray} \tag{6.22}$$

One finds that the required pan or yaw angle is simply

$$pan = \psi_{ray} \tag{6.23}$$

and the required tilt or pitch angle is

$$tilt = \theta_{ray} \tag{6.24}$$

Angular motion of this amount will bring the camera boresight onto the detected target. Once the pan and tilt unit bring the boresight of the camera in alignment with the object of interest, a coaligned ranging laser will similarly now have its axis aligned with the object of interest. See Figure 6.17.

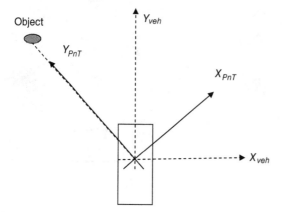

Figure 6.17 Pan and tilt unit rotated to required yaw and pitch for object of interest to appear at camera boresight.

The distance to the object of interest will now be measured with the ranging laser. With the boresight or the x axis of the sensor aimed at the object of interest, the position of this target expressed in the sensor frame will be given by

$$
\begin{bmatrix} X_{TS} \\ Y_{TS} \\ Z_{TS} \end{bmatrix}_{pointed} = \begin{bmatrix} 0 \\ Range \\ 0 \end{bmatrix} \tag{6.25}
$$

Converting this back to the coordinates of the sensor before the pan and tilt motion, we have

$$
\begin{bmatrix} X_{TS} \\ Y_{TS} \\ Z_{TS} \end{bmatrix}_{original} = \begin{bmatrix} -Range \cos\theta_{ray} \sin\psi_{ray} \\ Range \cos\theta_{ray} \cos\psi_{ray} \\ Range \sin\theta_{ray} \end{bmatrix} \tag{6.26}
$$

Several coordinate transformations are required to convert this vector to earth coordinates. The general rotation matrix presented earlier is repeated here

$$
R(\psi,\theta,\phi) = \begin{bmatrix} \cos\psi\cos\phi - \sin\psi\sin\theta\sin\phi & -\sin\psi\cos\theta & \cos\psi\sin\phi + \sin\psi\sin\theta\cos\phi \\ \sin\psi\cos\phi + \cos\psi\sin\theta\sin\phi & \cos\psi\cos\theta & \sin\psi\sin\phi - \cos\psi\sin\theta\cos\phi \\ -\cos\theta\sin\phi & \sin\theta & \cos\theta\cos\phi \end{bmatrix} \tag{6.27}
$$

The rotation matrix for the sensor with respect to vehicle is given by

$$
R_{SV} = \begin{bmatrix} \cos\psi_{SV} & -\sin\psi_{SV}\cos\theta_{SV} & \sin\psi_{SV}\sin\theta_{SV} \\ \sin\psi_{SV} & \cos\psi_{SV}\cos\theta_{SV} & -\cos\psi_{SV}\sin\theta_{SV} \\ 0 & \sin\theta_{SV} & \cos\theta_{SV} \end{bmatrix} \tag{6.28}
$$

These represent the orientation of the platform of the pan and tilt unit with respect to the vehicle under zero pan and tilt. It is simpler than equation (6.27) because roll of the pan and tilt unit is zero. These also correspond to the camera orientation with respect to the vehicle because it has been assumed that the camera is mounted with zero yaw, pitch, and roll with respect to the pan and tilt platform.

The rotation matrix for the vehicle with respect to earth is

$$
R_{VE} = \begin{bmatrix} \cos\psi_{VE}\cos\phi_{VE} - & & \cos\psi_{VE}\sin\phi_{VE} + \\ \sin\psi_{VE}\sin\theta_{VE}\sin\phi_{VE} & -\sin\psi_{VE}\cos\theta_{VE} & \sin\psi_{VE}\sin\theta_{VE}\cos\phi_{VE} \\ \sin\psi_{VE}\cos\phi_{VE} + & & \sin\psi_{VE}\sin\phi_{VE} - \\ \cos\psi_{VE}\sin\theta_{VE}\sin\phi_{VE} & \cos\psi_{VE}\cos\theta_{VE} & \cos\psi_{VE}\sin\theta_{VE}\cos\phi_{VE} \\ -\cos\theta_{VE}\sin\phi & \sin\theta_{VE} & \cos\theta_{VE}\cos\phi_{VE} \end{bmatrix}
$$

$$(6.29)$$

The yaw angle (ψ_{VE}), pitch angle (θ_{VE}), and roll angle (ϕ_{VE}), all vary and are read out from the robot attitude sensor.

Now the target location is converted to vehicle coordinates. The required equation is

$$
\begin{bmatrix} X_T \\ Y_T \\ Z_T \end{bmatrix}_{vehicleo\ coords} = [R_{SV}] \begin{bmatrix} X_T \\ Y_T \\ Z_T \end{bmatrix}_{sensor\ coords} + \begin{bmatrix} X_S \\ Y_S \\ Z_S \end{bmatrix}_{vehicle\ coords} \qquad (6.30)
$$

This can be written more concisely as a single operation using the *homogeneous transformation* introduced earlier.

$$
\begin{bmatrix} X_T \\ Y_T \\ Z_T \\ 1 \end{bmatrix}_{vehicle\ coords} = [A_{SV}] \begin{bmatrix} X_T \\ Y_T \\ Z_T \\ 1 \end{bmatrix}_{sensor\ coords} \qquad (6.31)
$$

where for A_{SV} we have

$$
A_{SV} = \begin{bmatrix} & & & X_{S\ veh\ coords} \\ & R_{SV} & & Y_{S\ veh\ coords} \\ & & & Z_{S\ veh\ coords} \\ 0 & 0 & 0 & 1 \end{bmatrix} \qquad (6.32)
$$

The fourth column represents the position of the origin of the sensor frame in vehicle coordinates, i.e., the dimensions associated with the location of the sensor on the vehicle. They are fixed quantities and are measured with respect to the position of the DGPS receiver, since this is where the origin of the vehicle frame is defined to be.

For A_{VE} we have

$$A_{VE} = \begin{bmatrix} & & & X_{V \, earthoords} \\ & R_{VE} & & Y_{V \, earthoords} \\ & & & Z_{V \, earthoords} \\ 0 & 0 & 0 & 1 \end{bmatrix} \tag{6.33}$$

where the fourth column represents the position of the origin of the vehicle frame in earth coordinates, i.e., the coordinates obtained from DGPS receiver. Using these homogeneous transformation matrices the position of the object of interest in earth axes will be given by

$$\begin{bmatrix} X_T \\ Y_T \\ Z_T \\ 1 \end{bmatrix}_{earth\,coords} = \begin{bmatrix} & & & X_{veh\,earth\,coords} \\ & R_{VE} & & Y_{veh\,earth\,coords} \\ & & & Z_{veh\,earth\,coords} \\ 0 & 0 & 0 & 1 \end{bmatrix}$$

$$\begin{bmatrix} & & & X_{S\,vehicle\,coords} \\ & R_{SV} & & Y_{S\,vehicle\,coords} \\ & & & Z_{S\,vehicle\,Coords} \\ 0 & 0 & 0 & 1 \end{bmatrix} \begin{bmatrix} X_T \\ Y_T \\ Z_T \\ 1 \end{bmatrix}_{sensor\,coords} \tag{6.34}$$

or

$$\begin{bmatrix} X_T \\ Y_T \\ Z_T \\ 1 \end{bmatrix}_{earth\,coords} = [A_{VE}][A_{SV}] \begin{bmatrix} X_T \\ Y_T \\ Z_T \\ 1 \end{bmatrix}_{sensor\,coords} \tag{6.35}$$

In terms of the nonhomogeneous representation, the above equations for the target in earth coordinates are equivalent to

$$\begin{bmatrix} X_T \\ Y_T \\ Z_T \end{bmatrix}_{earth\,coords} = \begin{bmatrix} X_V \\ Y_V \\ Z \end{bmatrix}_{earth\,coords} + R_{VE} \begin{bmatrix} X_S \\ Y_S \\ Z_S \end{bmatrix}_{vehicle\,coords}$$

$$+ R_{VE}R_{SV} \begin{bmatrix} -Range \cos\theta_{ray} \sin\psi_{ray} \\ Range \cos\theta_{ray} \cos\psi_{ray} \\ Range \sin\theta_{ray} \end{bmatrix} \tag{6.36}$$

Although less concise than when using the homogeneous transformation matrices, this last form is intuitively appealing since it is the sum of three recognizable terms: (a) position of the vehicle with respect to the earth in earth coordinates plus (b) position of the sensor with respect to the vehicle converted to earth coordinates plus (c) position of the target with respect to the sensor converted to vehicle coordinates and then to earth coordinates. Here, the rotation matrix (R_{SV}) is evaluated at zero pan and tilt, the conditions that existed at the time of the detection.

The above analysis has been based on the assumption that the target has been detected by the camera with the pan and tilt platform at zero pan and zero tilt. The required pan and tilt for aligning the camera boresight and the ranging device with the target were then computed based on the camera pixel coordinates of the target. After measuring the range to the target, the target location was then determined in earth coordinates.

6.6 Detection Sensor in Scanning Mode

A more likely scenario would be that the pan and tilt unit is used in a scanning mode and that the target was detected when the pan and tilt coordinates were not zero. In Figure 6.18a and b, IR images with different pan angles are shown.

The question then is what the new pan and tilt coordinates should be in order to align the camera boresight (and thus the laser boresight) with the target for measuring the range. See Figure 6.19.

The unit vector at detection in camera coordinates would be given by

$$X_{T\ sensor\ a\ coords} = \begin{bmatrix} -\sin\psi_{ray}\cos\theta_{ray} \\ \cos\psi_{ray}\cos\theta_{ray} \\ \sin\theta_{ray} \end{bmatrix} \tag{6.37}$$

(a)   (b)

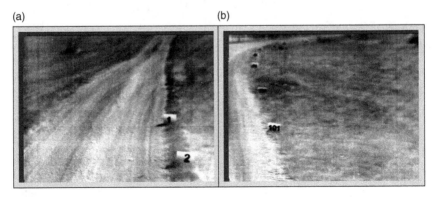

Figure 6.18 (a) Camera pointed straight ahead. (b) Camera panned to the right.

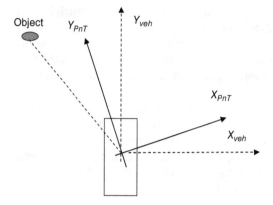

The rotation of the camera with respect to the vehicle is given by the current pan and tilt angles at the time of target detection. This rotation matrix is

$$R_{pan\ \&\ tilt} = \begin{bmatrix} \cos{(pan_1)} & -\sin{(pan_1)} & 0 \\ \sin{(pan_1)} & \cos{(pan_1)} & 0 \\ 0 & 0 & 1 \end{bmatrix} \begin{bmatrix} 1 & 0 & 0 \\ 0 & \cos{(tilt_1)} & -\sin{(tilt_1)} \\ 0 & \sin{(tilt_1)} & \cos{(tilt_1)} \end{bmatrix}$$

(6.38)

The unit ray to the target in vehicle coordinates is given by

$$X_{T\ vehicle\ coords} = \begin{bmatrix} \cos{(pan_1)} & -\sin{(pan_1)} & 0 \\ \sin{(pan_1)} & \cos{(pan_1)} & 0 \\ 0 & 0 & 1 \end{bmatrix}$$

$$\begin{bmatrix} 1 & 0 & 0 \\ 0 & \cos{(tilt_1)} & -\sin{(tilt_1)} \\ 0 & \sin{(tilt_1)} & \cos{(tilt_1)} \end{bmatrix} \begin{bmatrix} -\sin{\psi_{ray}}\cos{\theta_{ray}} \\ \cos{\psi_{ray}}\cos{\theta_{ray}} \\ \sin{\theta_{ray}} \end{bmatrix}$$

(6.39)

or

$$X_{T\ vehicle\ coords} = \begin{bmatrix} \cos{(pan_1)} & -\sin{(pan_1)}\cos{(tilt_1)} & \sin{(pan_1)}\sin{(tilt_1)} \\ \sin{(pan_1)} & \cos{(pan_1)}\cos{(tilt_1)} & -\cos{(pan_1)}\sin{(tilt_1)} \\ 0 & \sin{(tilt_1)} & \cos{(tilt_1)} \end{bmatrix}$$

$$\begin{bmatrix} -\sin{\psi_{ray}}\cos{\theta_{ray}} \\ \cos{\psi_{ray}}\cos{\theta_{ray}} \\ \sin{\theta_{ray}} \end{bmatrix}$$

(6.40)

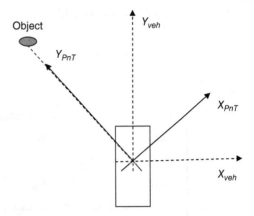

Figure 6.20 Pan and tilt unit rotated by required yaw and pitch to bring detected object to camera boresight.

In order to measure the range to the target, we desire the pan and tilt motion that will place the target at the boresight of the camera. See Figure 6.20. When this is accomplished, the expression for the unit ray in camera coordinates will be

$$X_{T \, sensora \, coords} = \begin{bmatrix} 0 \\ 1 \\ 0 \end{bmatrix} \qquad (6.41)$$

The equation for the unit ray in vehicle coordinates will then become

$$X_{T \, vehicle \, coords} = \begin{bmatrix} \cos(pan_2) & -\sin(pan_2) & 0 \\ \sin(pan_2) & \cos(pan_2) & 0 \\ 0 & 0 & 1 \end{bmatrix} \begin{bmatrix} 1 & 0 & 0 \\ 0 & \cos(tilt_2) & -\sin(tilt_2) \\ 0 & \sin(tilt_2) & \cos(tilt_2) \end{bmatrix} \begin{bmatrix} 0 \\ 1 \\ 0 \end{bmatrix} \qquad (6.42)$$

or

$$X_{T \, vehicle \, coords}$$
$$= \begin{bmatrix} \cos(pan_2) & -\sin(pan_2)\cos(tilt_2) & \sin(pan_2)\sin(tilt_2) \\ \sin(pan_2) & \cos(pan_2)\cos(tilt_2) & -\cos(pan_2)\sin(tilt_2) \\ 0 & \sin(tilt_2) & \cos(tilt_2) \end{bmatrix} \begin{bmatrix} 0 \\ 1 \\ 0 \end{bmatrix} \qquad (6.43)$$

Equating the two expressions for $X_{T\,vehicle\,coords}$ yields the vector equation

$$
\begin{bmatrix}
-\sin(pan_2)\cos(tilt_2) \\
\cos(pan_2)\cos(tilt_2) \\
\sin(tilt_2)
\end{bmatrix}
$$

$$
=
\begin{bmatrix}
-\cos(pan_1)\sin\psi_{ray}\cos\theta_{ray} - \sin(pan_1)\cos(tilt_1)\cos\psi_{ray}\cos\theta_{ray} \\
+ \sin(pan_1)\sin(tilt_1)\sin\theta_{ray} \\
-\sin(pan_1)\sin\psi_{ray}\cos\theta_{ray} + \cos(pan_1)\cos(tilt_1)\cos\psi_{ray}\cos\theta_{ray} \\
-\cos(pan_1)\sin(tilt_1)\sin\theta_{ray} \\
\sin(tilt_1)\cos\psi_{ray}\cos\theta_{ray} + \cos(tilt_1)\sin\theta_{ray}
\end{bmatrix}
$$

$$(6.44)$$

One can solve for pan_2 by equating the ratio of the negative of the first component to the second component on each side, i.e.,

$$
\tan(pan_2)
$$

$$
= \left\{ \frac{\cos(pan_1)\sin\psi_{ray}\cos\theta_{ray} + \sin(pan_1)\cos(tilt_1)\cos\psi_{ray}\cos\theta_{ray} - \sin(pan_1)\sin(tilt_1)\sin\theta_{ray}}{-\sin(pan_1)\sin\psi_{ray}\cos\theta_{ray} + \cos(pan_1)\cos(tilt_1)\cos\psi_{ray}\cos\theta_{ray} - \cos(pan_1)\sin(tilt_1)\sin\theta_{ray}} \right\}
$$

or

$$
pan_2 = \tan^{-1}
$$

$$
\left\{ \frac{\cos(pan_1)\sin\psi_{ray}\cos\theta_{ray} + \sin(pan_1)\cos(tilt_1)\cos\psi_{ray}\cos\theta_{ray} - \sin(pan_1)\sin(tilt_1)\sin\theta_{ray}}{-\sin(pan_1)\sin\psi_{ray}\cos\theta_{ray} + \cos(pan_1)\cos(tilt_1)\cos\psi_{ray}\cos\theta_{ray} - \cos(pan_1)\sin(tilt_1)\sin\theta_{ray}} \right\}
$$

$$(6.45)$$

One finds $tilt_2$ by equating the third components on each side, i.e.,

$$
\sin(tilt_2) = \sin(tilt_1)\cos\psi_{ray}\cos\theta_{ray} + \cos(tilt_1)\sin\theta_{ray}
$$

or

$$
tilt_2 = \sin^{-1}\left\{ \sin(tilt_1)\cos\psi_{ray}\cos\theta_{ray} + \cos(tilt_1)\sin\theta_{ray} \right\} \quad (6.46)
$$

After setting the pan and tilt unit to these angles, the longitudinal axis of the camera and range finder will be aligned with the target. Once the range has been measured, one proceeds as before to convert the measurement to earth coordinates.

Example 5 *For determining the required pan and tilt angles for bringing the boresight to the target, one might be tempted to simply add the azimuth of the ray pointing at the target to pan_1 and to add the elevation of the ray pointing at the target to $tilt_1$. Demonstrate numerically that this is not the correct approach.*

Solution 5

Let $pan_1 = 20$ and $\psi_{ray} = 7$. Also let $tilt_1 = 30$ and $\theta_{ray} = 5$. By adding pan_1 and ψ_{ray} one would conclude that $pan_2 = 27$ when in fact use of the correct procedure yields $pan_2 - 28.4962$. Similarly by adding $tilt_1$ and θ_{ray} one would conclude that $tilt_2 = 35$ when in fact use of the correct procedure yields $tilt_2 = 34.7407$.

Exercises

1   **A** A mobile robot is at location $x = 102$ m, $y = 59$ m, and $z = 1$ m when specified in a local coordinate frame. The attitude of the robot is yaw = 37°, pitch = 5°, and roll = 4°. An object of interest is sensed by a camera and a range sensor mounted on the robot. The object is determined to be located at $x = 3$ m, $y = 15$ m, and $z = -2.2$ m in the robot frame. Compute the location of the object in the local coordinate frame.

   B A second object is identified. Now the robot position is $x = 120$ m, $y = 40$ m, and $z = 0$ and its attitude is yaw = 40°, pitch = 0, and roll = 0. The object is determined to be located at $x = -7$ m, $y = 10$ m, and $z = 2$ m in the robot frame. Assume that the robot has differential wheel steering as well as an adjustable suspension system giving it the ability to vary its pitch by elevating the front or back with respect to the wheels. It is desired to aim the longitudinal axis of the robot at the object of interest. What are the required rotation angles for the robot?

2   **A** The robot has an IR camera mounted at its center with respect to width and length. The camera is 1 m above the ground and is pitched down at 12° below horizontal. The field of view is 14° in height and 18° in width. There are 320 pixels across and 240 pixels down. A target is detected in the 85th row and 203rd column. What are the pan and tilt commands to align the camera boresight with the target?

   B Assume that a ranging laser that is aligned with the camera boresight measures the range to the target as 15 m. Determine the location of the target. The robot GPS readings give its location as Easting = 755,295 and Northing = 3,685,240. The vehicle heading measured in the counterclockwise direction from the Y axis, i.e., from North, is −135°.

3 An IR camera has a vertical field of view of 12° and a horizontal field of view of 16°. It has 320 columns and 240 rows of pixels. It is mounted on a robot at a height of 2 m and a pitch of –6°. The x and y coordinates of the camera with respect to the vehicle are zero. Also yaw and roll of the camera with respect to the robot are zero. An object of interest is detected at the 30th row and the 48th column. Find the location of the object in vehicle coordinates.

4 A radar sensor mounted on a mobile robot shows an object of interest in the pixel of the 100th row and the 25th column. The field of view is laid out to cover the area beginning 10 m in front of the robot up to 50 m away. It extends over a width of 10 m. The dimension of this field of view in pixels is 200 rows and 50 columns. What is the location of the object of interest in robot coordinates?

5 The radar sensor described in Exercise 4 shows an object of interest in the 85th row and the 35th column. What is the location of the object of interest in robot coordinates? If the robot is located at $x = 300$ and $y = 200$ with a heading of 45°, all in a local coordinated system, what is the location of the object of interest in this local coordinate system?

6 A robot with a camera mounted on a pan and tilt unit detects an object when the pan angle is 25° and the tilt angle is –7°. The camera has a vertical field of view of 12° and a horizontal field of view of 16°. It has 320 columns and 240 rows of pixels. The object appears in the 112th row and the 143rd column. What should the pan and tilt angles be set to in order to bring the boresight of the camera onto the target for a range measurement?

7 Assume that the range measurement described in Exercise 6 turned out to be 75 m. What would be the location of the object in robot coordinates?

References

Bar-Shalom, Y. and Fortmann, T., *Tracking and Data Association*, Academic Press, 1988.

Cook, G., Sherbondy, K., and Jakkidi, S., "Geo-location of Mines Detected via Vehicular-Mounted Forward-Looking Sensors," *Proceedings of SPIE Conference on Detection and Remediation Technologies for Mine and Minelike Targets VI*, Vol. 4394 (pp. 922–933, Orlando, FL, April 16–20, 2001).

Jakkidi, S. and Cook, G., "Geo-location of Detected Landmines via Mobile Robots Equipped with Multiple Sensors," *Proceedings of IECON 2002* (pp. 1972–1977, Seville, Spain, November 4–8, 2002).

Jakkidi, S. and Cook, G., "Landmine Geo-location; Dynamic Modeling and Target Tracking," *IEEE Transactions on Aerospace and Electronics*, Vol. 41, No. 1 (January 2005), pp. 51–59.

Johnson, P. G. and Howard, P., "Performance Results of the EG&G Vehicle-Mounted Mine Detector," *Proceedings of SPIE, Detection and Remediation Technologies for Mines and Mine-like Targets IV* (pp. 1149–1159, Orlando, FL, April 5–9, 1999).

Kansal, S., Jakkidi, S., and Cook, G., "The Use of Mobile Robots for Remote Sensing and Object Localization," *Proceedings of IECON 2003* (pp. 279–284, Roanoke, VA, November 2–6, 2003).

Kositsky, J. and Milanfar, P., "Forward-Looking High Resolution GPR System," *Proceedings of SPIE, Detection and Remediation Technologies for Mines and Mine-like Targets IV* (pp. 1052–1062, Orlando, FL, April 5–9, 1999).

Lloyd, J. M., *Thermal Imaging Systems*, Springer, 1975.

Skolink, I. M., *Introduction to Radar Systems*, McGraw Hill, 1980.

Wu, Y. A., "EO Target Geolocation Determination," *Proceedings of IEEE Conference on Decision and Control* (pp. 2766–2771, December 1995).

7

Target Tracking Including Multiple Targets with Multiple Sensors

7.1 Introduction

This chapter is devoted to tracking the coordinates of detected objects of interest as the mobile robot moves along in its search. There can be multiple detections of a given sensor as well as detections of a given target by more than one sensor. All these detections can be used to improve the estimate of the coordinates of the objects of interest. In some cases, multiple targets may be detected and tracked. Means of associating measurements with the proper targets are also discussed.

7.2 Regions of Confidence for Sensors

Every measurement is accompanied by some uncertainty, which depends on the variance of the measurement error. Thus, when one computes the ground coordinates of an object of interest, there is associated with this value a region of confidence. The smaller the errors, the tighter will be the region of confidence. In the principal axes, the boundary of the region of confidence is described by the equation

$$\frac{(x - x_{est})^2}{\sigma_x^2} + \frac{(y - y_{est})^2}{\sigma_y^2} = \gamma \tag{7.1}$$

where σ_x and σ_y are the standard deviations of the measurement error in the x and y directions, respectively. It is assumed that the errors have Gaussian distributions. This boundary, which is an ellipse, contains the true target location with probability P given in Table 7.1.

The user selects γ in accordance with the certainty desired. To have a high degree of confidence that the object of interest is within the boundary, one selects a high value for P. The result is a large value for the associated γ and therefore a large ellipse. In the original coordinates the ellipse axes may be

Mobile Robots: Navigation, Control and Sensing, Surface Robots and AUVs,
Second Edition. Gerald Cook and Feitian Zhang.
© 2020 by The Institute of Electrical and Electronics Engineers, Inc.
Published 2020 by John Wiley & Sons, Inc.

Table 7.1 Typical values of gamma from chi-square tables.

P	0.5	0.90	0.95	0.99
γ	1.39	4.6	5.99	9.2

rotated, i.e., cross variance is possible. The variance of the measurement errors depends on the errors associated with the sensor itself as well as errors in the estimates of the vehicle position and attitude. In the original coordinates, the general equation of an ellipse defines the boundary of the confidence region and is given by

$$
\begin{bmatrix} x - x_{est} \\ y - y_{est} \end{bmatrix}^T \begin{bmatrix} \sigma_x^2 & \sigma_{xy} \\ \sigma_{yx} & \sigma_y^2 \end{bmatrix}^{-1} \begin{bmatrix} x - x_{est} \\ y - y_{est} \end{bmatrix} = \gamma \tag{7.2}
$$

or

$$
\begin{bmatrix} \Delta x \\ \Delta y \end{bmatrix}^T \begin{bmatrix} \sigma_x^2 & \sigma_{xy} \\ \sigma_{yx} & \sigma_y^2 \end{bmatrix}^{-1} \begin{bmatrix} \Delta x \\ \Delta y \end{bmatrix} = \gamma \tag{7.3}
$$

The following describes how to convert this to the simpler ellipse representation. Let the eigenvectors of the inverse of the covariance matrix be represented by ξ_1 and ξ_2, and define the matrix

$$
M = [\xi_1 \quad \xi_2]
$$

Then the above equation can be written as

$$
\begin{bmatrix} \Delta x \\ \Delta y \end{bmatrix}^T MM^{-1} \begin{bmatrix} \sigma_x^2 & \sigma_{xy} \\ \sigma_{yx} & \sigma_y^2 \end{bmatrix}^{-1} MM^{-1} \begin{bmatrix} \Delta x \\ \Delta y \end{bmatrix} = \gamma
$$

Now since the inverse of the covariance matrix is symmetric, its eigenvectors will be orthogonal. They may also be normalized yielding the result $M^{-1} = M^T$. The above equation may then be rewritten as

$$
\begin{bmatrix} \Delta x \\ \Delta y \end{bmatrix}^T MM^T \begin{bmatrix} \sigma_x^2 & \sigma_{xy} \\ \sigma_{yx} & \sigma_y^2 \end{bmatrix}^{-1} MM^T \begin{bmatrix} \Delta x \\ \Delta y \end{bmatrix} = \gamma \tag{7.4}
$$

or

$$
\begin{bmatrix} \Delta \hat{x} \\ \Delta \hat{y} \end{bmatrix}^T \begin{bmatrix} 1/\hat{\sigma}_x^2 & 0 \\ 0 & 1/\hat{\sigma}_y^2 \end{bmatrix} \begin{bmatrix} \Delta \hat{x} \\ \Delta \hat{y} \end{bmatrix} = \gamma \tag{7.5}
$$

where

$$\begin{bmatrix} \Delta \widehat{x} \\ \Delta \widehat{y} \end{bmatrix} = M^T \begin{bmatrix} \Delta x \\ \Delta y \end{bmatrix} \qquad (7.6)$$

and

$$\begin{bmatrix} 1/\widehat{\sigma}_x^2 & 0 \\ 0 & 1/\widehat{\sigma}_y^2 \end{bmatrix} = M^T \begin{bmatrix} \sigma_x^2 & \sigma_{xy} \\ \sigma_{yx} & \sigma_y^2 \end{bmatrix}^{-1} M \qquad (7.7)$$

It is seen that the above representation of the ellipse reduces to the simpler form

$$\frac{\Delta \widehat{x}^2}{\widehat{\sigma}_x^2} + \frac{\Delta \widehat{y}^2}{\widehat{\sigma}_y^2} = \gamma \qquad (7.8)$$

If desired, one may compute the coordinates of the ellipse in these normal coordinates and then map them into the original coordinates.

It has been noted that for objects sensed via cameras, one can directly determine the direction to the object but not the range. If a single camera is used, one can make the assumptions described in the foregoing and estimate the range. There are errors associated with this estimate that depend on errors in measurement of vehicle attitude and unevenness of terrain. Thus, the uncertainty in knowledge of the y (or down-range) component can be quite large whereas the x (or cross-range) component is more precisely known. If one constructs an ellipse of uncertainty about the measurement corresponding to the boundary within which the true position exists with a given probability, this ellipse is elongated in the down-range direction and more slender in the cross-range direction, reflecting the difference in error variance in the two coordinates.

In contrast, the radar sensor just described has very good precision in the down-range direction, limited only by the resolution of the timing circuitry, whereas its cross-range precision is poorer because of the limited horizontal extent of the array causing one to have to resolve the intersection of nearly parallel ellipses. If one constructs an ellipse of uncertainty about this measurement, it would be elongated in the cross-range direction and more slender in the down-range direction.

Combined measurements have the property that as more measurements are taken, the variance of the estimate monotonically decreases. In fact, their variances combine in the same way as do resistors in parallel. As an example, if one optimally estimates a scalar quantity, X from n independent measurements $Y_1, Y_2, ..., Y_n$ where the covariance of the error in each measurement is σ_i^2 and the mean is zero, then the estimate is given by

$$\hat{X} = \frac{1/\sigma_1^2}{1/\sigma_1^2 + 1/\sigma_2^2 + \cdots + 1/\sigma_n^2} Y_1 + \cdots + \frac{1/\sigma_n^2}{1/\sigma_1^2 + 1/\sigma_2^2 + \cdots + 1/\sigma_n^2} Y_n$$

(7.9)

The covariance of the error in the final estimate is

$$\sigma^2 = \frac{1}{1/\sigma_1^2 + 1/\sigma_2^2 + \cdots + 1/\sigma_n^2}$$

(7.10)

Thus for estimating the value of a fixed scalar quantity, the variance can never increase as additional measurements are taken. And as illustrated above, the variance for the estimate will never be greater than the smallest measurement variance. This scalar quantity could represent one component of a vector describing the location of a stationary object. Applying the above results to this situation, one has that the variance of the error in any direction can never increase and further that the variance in that direction would be no greater than the smallest measurement variance in that direction. Because radar is so precise in the down-range coordinate, which is where IR is the worst, and the IR is better in the cross-range direction where this particular type of radar is the worst, the combination of radar and IR sensing offers great promise for ground registration. These two sensors may be said to have orthogonal axes of precision. This is a fortuitous set of circumstances that one may exploit.

A diagram illustrating the components of sensor measurement error is shown in Figure 7.1.

The variances for the down-range and cross-range components of this error are given by

$$\sigma_{downrange}^2 = \sigma_{dnrng-sen}^2$$

(7.11)

$$\sigma_{crossrange}^2 = \sigma_{crsrng-sen}^2 + r^2 \sigma_{\psi-veh}^2$$

(7.12)

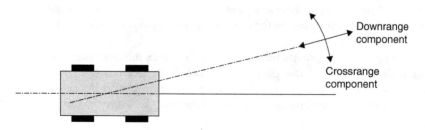

Figure 7.1 Illustration of components of measurement error.

The covariance matrix describing this uncertainty in earth coordinates is given by

$$\begin{bmatrix} R_{XX} & R_{XY} \\ R_{YX} & R_{YY} \end{bmatrix} = \begin{bmatrix} \sigma^2_{x-veh} & \sigma_{xy-veh} \\ \sigma_{yx-veh} & \sigma^2_{y-veh} \end{bmatrix} + R^{earth}_{veh} \begin{bmatrix} \sigma^2_{downrange} & 0 \\ 0 & \sigma^2_{crossrange} \end{bmatrix}$$

(7.13)

where σ_{x-veh} and σ_{y-veh} are obtained from the Kalman Filter associated with the vehicle model. This covariance matrix is processed to yield the required coordinate transformation and the final σ^2_x and σ^2_y to be used in equation (7.8). Greater values of σ yield larger ellipses. The eccentricity of the ellipse is determined by the relative values of σ_x and σ_y.

Based on the algorithm used for estimating the range to the object of interest, one can determine that the covariance for the camera sensor in the down-range coordinate is

$$\sigma^2_{dnrng-sen} = \left(\frac{r^2}{H}\right)^2 \sigma^2_{\theta-veh}$$

(7.14)

Note that this quantity becomes very large as r increases. Sensor error variance in the cross-range direction is limited only by pixel resolution. Using typical values for the various parameters, the ellipse describing the region of confidence is shown in Figures 7.2 and 7.3.

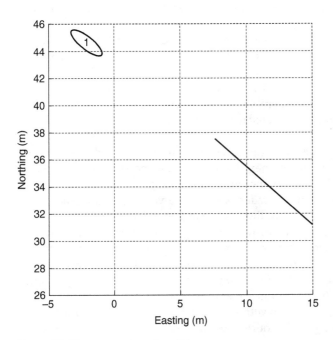

Figure 7.2 IR sensor region of confidence.

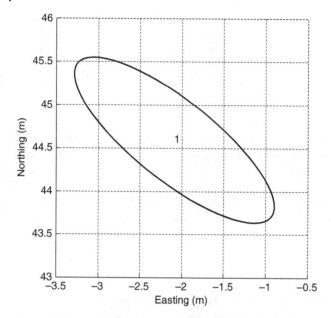

Figure 7.3 Expanded view of the region of confidence for IR.

In Figure 7.2, the path of the mobile robot toward the object of interest is also shown. The large value of error variance in the down-range direction compared to that in the cross-range direction is apparent. In contrast to the IR sensor, the error variance for the radar considered here in the down-range direction is small but it is larger in the cross direction. Using typical values, the ellipse describing the region of confidence for radar is shown in Figures 7.4 and 7.5.

The results shown were generated using simulated IR and Radar object detections. The parameter γ was selected to be 4.6 corresponding to a 90% confidence probability.

Figures 7.6 and 7.7 illustrate the successive ellipses obtained when the two sensors are combined. In the next figure, the first measurement is from an IR sensor and has a large covariance in the down-range coordinate. The second measurement comes from the radar, which has a very tight covariance in the down-range coordinate. Even though radar has a large covariance in the cross-range coordinate, this dimension has already been tightened by the preceding IR measurement. Thus, the cumulative confidence region after processing both the measurements has good behavior in both the down-range and cross-range coordinates. The succeeding figure illustrates the same phenomenon with the order of IR and radar measurements reversed.

The use of multiple sensors with complementary precision characteristics for determining the ground coordinates of detected objects of interest has

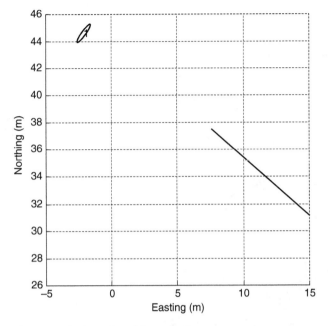

Figure 7.4 Region of confidence for linear array radar sensor.

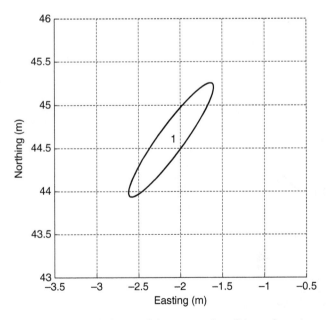

Figure 7.5 Expanded view of the region of confidence for radar.

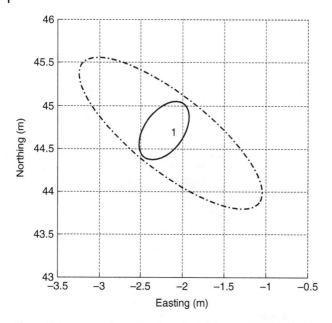

Figure 7.6 Convergence of region of confidence with combination of IR (dashed) followed by radar (solid) sensors.

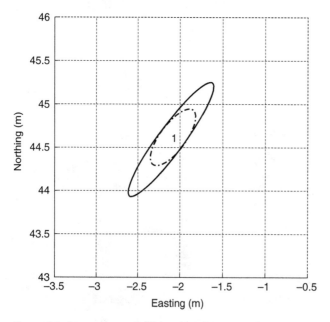

Figure 7.7 Convergence of region of confidence with radar (solid) followed by IR (dashed) sensors.

been demonstrated. Here an IR sensor that has good precision in the cross-range direction is fused with a radar sensor that has good precision in the down-range direction. This fusion of sensors whose axes of highest precision are orthogonal to each other exploits the best characteristics of each sensor and provides dramatic improvement in ground registration precision in both directions, making it a very effective ground registration combination.

7.3   Model of Target Location

Section 7.1 addressed the problem of measuring the location of an object of interest, or target, utilizing various sensors. As was discussed, in some cases one may have multiple sensors and may detect the same target with more than one sensor. It is also possible that the same target may be detected several times by a single sensor as the robot moves along. The goal of this section is to combine multiple measurements of the location of a target into an optimal estimate. Having a precise estimate of the location can be useful for accomplishing the next goal, which may be close-up examination, retrieval, or referral to another robot that would perform one of these tasks.

In order to combine multiple measurements from a single sensor or measurements from multiple sensors into a single estimate of the target location, one may use the Kalman Filter. Of course, central to the application of the Kalman Filter is a model of the process being observed. Here, the location of the target is sought, and we will make the assumption that it is stationary. Thus a simple model for the position coordinates of the ith target is

$$
\begin{aligned}
x_i(k+1) &= x_i(k) \\
y_i(k+1) &= y_i(k)
\end{aligned}
\tag{7.15}
$$

When the measurement of the location of the detected target is converted to earth coordinates, one component of the measurement is the estimated location and attitude of the robot itself. The covariances of the errors in these estimates contribute to the total measurement error of the location of the target and they must be included in the analysis. While the sensor measurement noise itself is modeled as being "white," the errors in the estimates of the robot position and attitude are in fact colored; these estimates having been obtained via a dynamic model and the use of the Kalman Filter. Thus the errors in measurement of the earth coordinates of the targets have a component that is colored. The most accurate way of dealing with this situation would be to develop a colored noise model and incorporate it into the signal processing. See Jakkidi and Cook. Fortunately, the impact of the noise being colored is very small here and it can be treated as white without serious consequences.

The measurement equations for the ith target are as shown below:

$$
\begin{bmatrix} x \\ y \end{bmatrix}_{object\ i-earth\ coords} = \begin{bmatrix} x \\ y \end{bmatrix}_{vehicle-earth\ coords} + \begin{bmatrix} v_1 \\ v_2 \end{bmatrix}
$$
$$
+ R_{ve} \left(\begin{bmatrix} x \\ y \end{bmatrix}_{object\ i-vehicle\ coords} + \begin{bmatrix} n_{i1} \\ n_{i2} \end{bmatrix} \right)
$$

(7.16)

Note that the left side of the equation represents the x–y location of the target in earth coordinates. The first term on the right side is the x–y location of the vehicle in earth coordinates and the third term on the right side is the location of the target with respect to the robot in robot coordinates and then converted to earth coordinates.

The Kalman Filter equations for the objects of interest are quite simple. For the ith target we have for the prediction step

$$
\begin{bmatrix} \hat{x}_{k+1/k} \\ \hat{y}_{k+1/k} \end{bmatrix} = \begin{bmatrix} \hat{x}_{k/k} \\ \hat{y}_{k/k} \end{bmatrix} + \begin{bmatrix} w_1(k) \\ w_2(k) \end{bmatrix}
$$

(7.17)

and for the estimation step

$$
\begin{bmatrix} \hat{x}_{k+1/k+1} \\ \hat{y}_{k+1/k+1} \end{bmatrix} = \begin{bmatrix} \hat{x}_{k+1/k} \\ \hat{y}_{k+1/k} \end{bmatrix} + K \begin{bmatrix} x_{k+1\ meas} - \hat{x}_{k+1/k} \\ y_{k+1\ meas} - \hat{y}_{k+1/k} \end{bmatrix}
$$

(7.18)

The discrete-time model coefficient matrices used for computing the covariances and filter gain are

$$
A = \begin{bmatrix} 1 & 0 \\ 0 & 1 \end{bmatrix}
$$

(7.19)

$$
H = \begin{bmatrix} 1 & 0 \\ 0 & 1 \end{bmatrix}
$$

(7.20)

and

$$
G = \begin{bmatrix} 1 & 0 \\ 0 & 1 \end{bmatrix}
$$

(7.21)

The measurement covariance R would be similar to what was discussed in Chapter 6, depending on the sensor used. It represents the combined covariance of the sensor measurement noise and the error in vehicle position, all expressed in earth coordinates. For simplicity, it is approximated here as being constant.

Recall from the discussion in Chapter 5 that the steady-state gain of the filter diminishes when Q is small. For this reason, one normally includes a disturbance in the model. Nevertheless, even in the absence of the disturbance in the model, during the transient the gain is nonzero. Because of this and the interesting results that accrue, in the proceeding we shall explore the behavior of the filter for this particular problem with Q set to zero.

Multiple detections of the same object provide the opportunity to improve the estimate of the location of the object. This can be seen by the reduction in the norm of the covariance matrix as more detections are made. Recalling the equation for the covariances and gain, and using the definitions of A, G, H, Q, and R from above we have

$$P(k + 1/k) = P(k/k)$$

the covariance of the error in the prediction,

$$K(k) = P(k/k)[P(k/k) + R(k)]^{-1}$$

the gain of the filter, and

$$P(k + 1/k + 1) = [I - K(k)]P(k + 1/k) = [I - K(k)]P(k/k)$$

the covariance of the next estimate. Note that the possibility of a time varying measurement covariance has been included, i.e., R may be a function of k.

The above may be re-arranged to

$$P(k + 1/k + 1) = \left[(P(k/k) + R(k))(P(k/k) + R(k))^{-1} - P(k/k)(P(k/k) + R(k))^{-1}\right]P(k/k)$$

or

$$P(k + 1/k + 1) = R(k)[P(k/k) + R(k)]^{-1}P(k/k)$$

Now inverting each side of this equation yields

$$P(k + 1/k + 1)^{-1} = P(k/k)^{-1}[P(k/k) + R(k)]R(k)^{-1} = R(k)^{-1} + P(k/k)^{-1}$$

or

$$P(k + 1/k + 1) = \left(R(k)^{-1} + P(k/k)^{-1}\right)^{-1} \tag{7.22}$$

This illustrates that the Kalman Filter algorithm reduces to a familiar result from probability and statistics for the special case where both the A matrix and the H matrix are identity matrices. It is apparent from the above that $P(k/k)$, the covariance of the current estimate, decreases monotonically as new measurements are taken. Compare this equation with the scalar case of equation (7.10).

The equations for incorporating the new measurement into the next estimate also simplify for this case. Using

$$\hat{x}_{k+1/k+1} = \hat{x}_{k+1/k} + K_{k+1}\left(y_{k+1} - \hat{x}_{k+1/k}\right)$$

and realizing that

$$\hat{x}_{k+1/k} = A\hat{x}_{k/k} = \hat{x}_{k/k}$$

and that

$$K(k) = P(k/k)[P(k/k) + R(k)]^{-1}$$

the estimate becomes

$$\hat{x}_{k+1/k+1} = \hat{x}_{k/k} + P(k/k)[P(k/k) + R(k)]^{-1}\left(y_{k+1} - \hat{x}_{k/k}\right)$$

or

$$\hat{x}_{k+1/k+1} = (P(k/k) + R(k))(P(k/k) + R(k))^{-1}\hat{x}_{k/k} + P(k/k)[P(k/k) + R(k)]^{-1}\left(y_{k+1} - \hat{x}_{k/k}\right)$$

or

$$\hat{x}_{k+1/k+1} = R(k)[P(k/k) + R(k)]^{-1}\hat{x}_{k/k} + P(k/k)[P(k/k) + R(k)]^{-1}y_{k+1}$$

which may be re-written as

$$\hat{x}_{k+1/k+1} = \left[P(k/k)^{-1} + R(k)^{-1}\right]^{-1}P(k/k)^{-1}\hat{x}_{k/k} + \left[P(k/k)^{-1} + R(k)^{-1}\right]^{-1}R(k)^{-1}y_{k+1}$$

$$(7.23)$$

i.e., each term on the right is weighted according to the inverse of its associated covariance. Compare this result with the scalar case of equation (7.9).

Example 1 *A constant vector X is to be estimated from a series of measurements Y where*

$$Y(k) = X + v(k)$$

The noise sequence is independent with zero mean and has covariance R(k). Determine the estimate for X and the associated covariance of the error via processing the entire batch of data simultaneously.

Solution 1

Through successive use of the recursive relationships just presented, one can obtain as the estimate

$$\hat{X}_{N/N} = \left[R(1)^{-1} + R(2)^{-1} + \cdots + R(N)^{-1} \right]^{-1} \left[R(1)^{-1} Y(1) + R(2)^{-1} Y(2) + \cdots + R(N)^{-1} Y(N) \right]$$

with covariance of this estimate being

$$P(N/N) = \left[R(1)^{-1} + R(2)^{-1} + \cdots + R(N)^{-1} \right]^{-1}$$

Figure 7.8 presents target tracking results involving multiple detections of targets. The path of the robot as it moves from the lower right to the upper left can be seen from the figure. Ahead of the robot and to its left are two objects of interest and one object that was passed by the field of view of the sensor after the fourth measurement was taken. The series of ellipses illustrate the 0.9 confidence regions about the estimated locations of the targets. It is apparent that these regions shrink rapidly as successive measurements are made.

7.4 Inventory of Detected Targets

With the possibility of multiple targets in the field of view, the amount of information to be saved increases, resulting in the need for some sort of book-keeping technique. Book-keeping includes all the information pertaining to

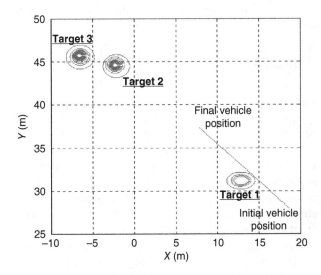

Figure 7.8 Estimated vehicle path with targets in field of view.

objects being tracked, current estimate of coordinates, and associated covariance. To facilitate the book-keeping, an inventory is developed. The approach used here is to have the inventory comprised of a matrix with each column representing the profile of an individual target. The columns are appended in the order of the appearance of the target (column 1 corresponds to the first target detected, column 2 corresponds to the second target detected, and so forth) and are updated as successive measurements are processed.

When a sensor measurement for the first target that has been detected is received, a ground location based on the sensor measurement is computed. This measurement is stored in the inventory. As successive sensor measurements are received, the inventory enables one to determine whether the measurement has arisen from a target previously detected or from a new target. This classification of targets in a new measurement frame, i.e., data association, is carried out by checking their computed locations against those of the targets that are already in the inventory. This process is called a "gating" operation. Clearly, it would make no sense to fuse measurement data originating from different targets. Hence, it is crucial that the data association operation be effectively carried out. Here two sources of uncertainty exist: the uncertainty of the location of the previously detected object represented by its associated estimate covariance listed in the inventory, as well as the uncertainty of the current measurement represented by its associated measurement covariance. If the error between a previously detected target and the current measurement falls within a specified confidence region, then the measurement is declared to have arisen from this previously detected target and is used to update the profile for that particular target. This includes updating the position estimate and the covariance. A measurement can be checked against the ith target by seeing if the inequality below is satisfied. Here it is seen that the combined covariance discussed above is used for the calculation required in this gating operation.

$$
\begin{bmatrix} x_{meas} - \hat{x}_{target\ i} \\ y_{meas} - \hat{y}_{target\ i} \end{bmatrix}^T \left(R_{meas} + P_{target\ i} \right)^{-1} \begin{bmatrix} x_{meas} - \hat{x}_{target\ i} \\ y_{meas} - \hat{y}_{target\ i} \end{bmatrix} \leq \gamma \qquad (7.24)
$$

Choosing the best value for the probability P (and thus the associated value for γ) may present a challenge. If one sets P as large as 0.99 (which corresponds to $\gamma = 9.2$), then 99% of the detections which are from a given target will get associated with that target; however, this corresponds to such a large ellipse that detections from other targets may also get associated with that target. Further, some detections may satisfy the gating inequality for more than one target in the inventory. One approach here would be to determine all the targets in the inventory that meet the threshold for a given detection and from these select that target from the inventory that is closest, i.e., that yields the smallest value for the LHS of inequality (7.24) to associate with the new detection. If there are multiple detections at a given time instant, one could require that only one detection be associated with a given target from the inventory at a given time instant and use the closest one. In contrast, one could use all of those detections satisfying the threshold for a given target in the inventory and weigh them according to their closeness to the target (see Bar Shalom). A simpler approach would be to use a smaller value for P at the risk of missing a few detections in order to prevent the incorrect association of detections from other targets. The final solution requires a compromise and depends on the density of targets as well as the measurement covariances.

If no target in the inventory yields an error that falls within this confidence region for a given detection, then the source of the measurement is declared to be a new target and a profile for it is initiated resulting in the inventory growing by one column.

Some examples of the inventory follow. It can be seen from Table 7.2 that when the object of interest was first detected, its location based on the sensor measurement was computed and stored in the Inventory as Target 1. The target was in the field of view until the fourth measurement was taken, i.e., time = 4. Therefore, the profile of Target 1 was updated three times. When the fourth measurement was taken, a second object of interest was detected in the sensor's field of view. The location of this target was computed and checked against the profile of the only target stored in the inventory. When the previously detected target did not fall within 0.9 confidence region surrounding the current measurement, a new profile was created in inventory under Target 2. This is shown in Table 7.3. It can also be seen from Table 7.3, that Target 1 disappeared after the 4th sensor measurement was taken and was not detected again and that Target 2 was still visible at the 35th sensor measurement.

Figure 7.9 further illustrates the tracking of these two targets. The current estimates of the coordinates of the targets are shown along with the extremities of the regions of confidence. The lack of measurements for Target 1 after the 4th sample is apparent as is the presence of measurements for Target 2 through 35 samples.

Table 7.2 Inventory with a single target in the field of view of sensor.

	Target 1
X coordinate	12.89
Y coordinate	31.16
Covariance 11	0.163
Covariance 12	0
Covariance 21	0
Covariance 22	0.047
Time	4

Table 7.3 Inventory with two targets in the field of view of sensor.

	Target 1	Target 2
X coordinate	12.89	−6.60
Y coordinate	31.16	47.75
Covariance 11	0.163	0.019
Covariance 12	0	0
Covariance 21	0	0
Covariance 22	0.047	0.01
Time	4	35

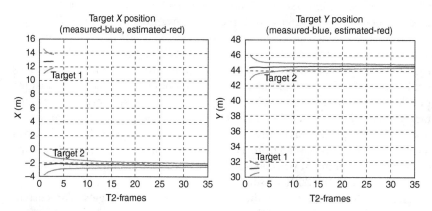

Figure 7.9 X and Y position of target in the field of view along with extremities of the regions of confidence.

The inventory also has the ability to handle the disappearance and reappearance of targets in the sensor field of view. Appearance of a target implies that an object of interest is present in the sensor field of view and disappearance of a target implies that the object of interest is absent from the sensor field of view. The profile of a target that has been detected is updated until the target disappears. The most recent estimates of the targets' locations as well as the covariance of error and other features are retained at the time of dropout. When the target reappears, the estimation process is resumed by picking up from estimates that existed at the time the target was last seen without any loss of information. The profile of the target is again updated accordingly.

Figure 7.10 shows such an example. It can be seen from this figure that Target 1 was detected in the sensor field of view in frame 1. The profile for that target was then initiated. When the 10th measurement was taken by the sensor,

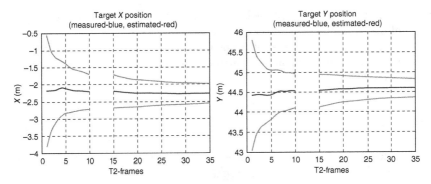

Figure 7.10 *X* and *Y* position of target in the field of view along with extremities of the regions of confidence, target disappears from Frames 10–15.

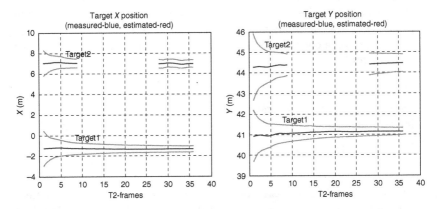

Figure 7.11 *X* and *Y* position of target in the field of view along with extremities of the regions of confidence.

Target 1 disappeared from the field of view. When the 15th measurement was taken by the sensor, Target 1 reappeared in the field of view. The estimation process resumed by using the latest information pertaining to Target 1 in the Inventory, and the estimate of the location of this object was further refined.

Figure 7.11 shows another example of a target dropout. It can be seen from this figure that initially both Target 1 and Target 2 were detected in the sensor field of view. The profiles of the objects were updated as new measurements were taken. When the ninth measurement of the targets was taken by the sensor, Target 2 disappeared but Target 1 remained continuously in the field of view of the sensor. The profile of Target 2 was retained in the Inventory at the time of dropout. When the 28th measurement was taken by the sensor, Target 2 reappeared in the sensor's field of view. The estimation process resumed by using the latest information in the Inventory pertaining to Target 2, and the estimate of the location of this object was further refined.

In all of the above examples, it is apparent that the regions of convergence shrink rapidly as successive measurements are made and the data are fused with the previous data.

Exercises

1   The location of an object of interest has been estimated to be $x = 23$ and $y = 8$. The associated covariance matrix is

$$R = \begin{bmatrix} 0.09 & 0.01 \\ 0.01 & 0.04 \end{bmatrix}$$

Determine the ellipse bounding the region of confidence for probability $P = 0.9$. Plot this ellipse in the original x–y coordinates. Note that the ellipse should be centered about the estimate. If desired, you may first plot the ellipse in normal coordinates and then map each point over to the original coordinates.

2   A mobile robot detects an object of interest at time $= 1$. Its local ground measurements are $x = 12$ and $y = 21$. The measurement covariance including all components robot position uncertainty and sensor error is

$$R_{meas} = \begin{bmatrix} 0.04 & 0.01 \\ 0.01 & 0.05 \end{bmatrix}$$

At time $= 2$, a detection is made at $x = 16$ and $y = 18$. The measurement covariance is the same as for the first detection. At time $= 3$, a detection is made at $x = 12.2$ and $y = 20.75$. The measurement covariance is again the same as for the previous detections. Show the Inventory for the files after time $= 1$, after time $= 2$, and after time $= 3$. The inventory should list the targets detected up until that time, the estimates of their positions, and the associated covariances. For testing to determine if a new detection is associated with a target being tracked, use an ellipse corresponding to $P = 0.9$ and add the covariance of the target estimate to the covariance of the detection measurement. Then check to see if inequality (7.23) is satisfied. An easy way to compute the covariance associated with an estimate based on multiple measurements is to use

$$P_{estimate} = \left(R_{meas1}^{-1} + R_{meas2}^{-1} + \cdots \right)^{-1}$$

3   Two objects of interest are being tracked. Their estimated locations are $\hat{x}_{target1} = 25$, $\hat{y}_{target1} = 45$, and $\hat{y}_{target1} = 45$, $\hat{y}_{target2} = 55$. The covariance

matrix for the estimated location of target 1 is $P_{target1} = \begin{bmatrix} 0.04 & 0.01 \\ 0.01 & 0.05 \end{bmatrix}$. For

target 2 it is $P_{target2} = \begin{bmatrix} 0.05 & 0.02 \\ 0.02 & 0.03 \end{bmatrix}$. A new measurement is taken with a

sensor for which the measurement noise has covariance

$R_{meas} = \begin{bmatrix} 0.1 & 0 \\ 0 & 0.1 \end{bmatrix}$. The sensor measurement is $x_{meas} = 29$, $y_{meas} = 41$.

Determine the target to associate this measurement with. Use as your criterion the one which yields the smallest value for the left side of the inequality (7.24).

4 Compute and plot the ellipses corresponding to the target covariances given in Exercise 3. Use the value for γ corresponding to $P = 0.9$.

5 Compute and plot the ellipses corresponding to measurement covariances $R_{meas1} = \begin{bmatrix} 0.5 & 0 \\ 0 & 0.1 \end{bmatrix}$ and $R_{meas2} = \begin{bmatrix} 0.1 & 0 \\ 0 & 0.5 \end{bmatrix}$. Use the value for γ corresponding to $P = 0.9$. Compare the two plots. Do the shapes correspond to any sensors discussed in the previous chapter?

References

Bar-Shalom, Y. and Fortmann, T., *Tracking and Data Association*, Academic Press, 1988.

Breuers, M. G. J., Schwering, P. B. W., and van de Brock, S. P., "Sensor Fusion Algorithms for the Detection of Land mines," *Detection and Remediation Technologies for Mines and Minelike Targets IV*, Vol. **3710** (pp. 1160–1166, International Society for Optics and Photonics, 1999).

Bryson, A. E. and Ho, Y.-C., *Applied Optimal Control: Optimization, Estimation and Control*, Blaisdell, Waltham, MA, 1969.

Cook, G. and Kansal, S., "Exploitation of Complementary Sensors for Precise Ground Registration of Sensed Objects," *Proceedings of SPIE Fourth International Asia-Pacific Environmental Remote Sensing Symposium*, Vol. **5657** (pp. 20–29, Honolulu, HI, November 8–11, 2004).

Cook, G., Sherbondy, K., and Jakkidi, S., "Geo-location of Mines Detected via Vehicular-Mounted Forward-Looking Sensors," *Proceedings of SPIE Conference on Detection and Remediation Technologies for Mine and Minelike Targets VI*, Vol. **4394** (pp. 922–933, Orlando, FL, April 16–20, 2001).

Dana, M. P., "Registration: A Prerequisite for Multiple Sensor Tracking," In *Multitarget-Multisensor Tracking: Advanced Applications*, edited by Y. Bar-Shalom, Artech House, 1989.

Drummond, Oliver E., "Multiple Sensor Tracking with Multiple Frame, Probabilistic Data Association," *Signal and Data Processing of Small Targets,* Vol. **2561** (pp. 322–336, International Society for Optics and Photonics, 1995).

Gelb, A. (ed.), *Applied Optimal Estimation,* MIT Press, Cambridge, MA, 1974.

Indar, Bhatia, Vince, Diehl, Tim, Moore, Jay, Marble, Ky, Tran, and Steve, Bishop, "Sensor Data Fusion for Mine Detection from a Vehicle-Mounted System," *SPIE,* Vol. **4038**, 2000.

Jakkidi, S. and Cook, G., "Geo-location of Detected Landmines via Mobile Robots Equipped with Multiple Sensors," *Proceedings of IECON 2002* (pp. 1972–1977, Seville, Spain, November 4–8, 2002).

Jakkidi, S. and Cook, G., "Landmine Geo-location; Dynamic Modeling and Target Tracking," *IEEE Transactions on Aerospace and Electronics,* Vol. **41**, No. 1 (January 2005), pp. 51–59.

McFee, J. E., Aitken, V. C., Chesney, R., Das, R., and Russel, K.L., "Multisensor Vehicle-Mounted Teleoperated Mine Detector with Data Fusion." *Detection and Remediation Technologies for Mines and Minelike Targets III,* Vol. **3392** (pp. 1082–1093. International Society for Optics and Photonics, 1998).

Meditch, J. S., *Stochastic Optimal Linear Estimation and Control,* McGraw-Hill, New York, NY, 1969.

Reid, D. B., "An Algorithm for Tracking Multiple Targets," *IEEE Transactions on Automatic Control,* Vol. **AC-24**, No. 6 (December 1979), pp. 843–854.

Thomopoulos, S. C., Ramanarayana, R., and Bougoulias, D., "Optimal Decision Fusion in Multiple-Sensor Systems," *IEEE Transactions on Aerospace and Electronic Systems,* Vol. **AES-23**, No. 5 (1987), pp. 644–652.

Tjuatja, S., Bredow, J. W., Fung, A. K., and Mitchell, O. R., "Combined Sensor Approach to the Detection and Discrimination of Anti-personnel Mines," *Proceedings of SPIE, Detection and Remediation Technologies for Mines and Mine-like Targets IV* (pp. 864–874, Orlando, FL, April 5–9, 1999).

8

Obstacle Mapping and Its Application to Robot Navigation

8.1 Introduction

In the process of accomplishing a given task, the mobile robot may encounter obstacles other than objects of interest in its path. It is important to detect such obstacles before having a collision that could result in damage to the robot or its instrumentation. Thus, a variety of sensors may be operated as the robot moves along in order to provide advance warning of obstacles. The precision need not be as great as that required for geo-registration of objects of interest, but it must be sufficiently precise to permit avoidance of the obstacle by the mobile robot without requiring excessive spacing or miss-distance. Some of the techniques described for detection and geo-registration of objects of interest may also be used for detection and geo-registration of obstacles. Previously detected and geo-registered obstacles may in turn be used as an aid in navigation. This is especially important when the robot is isolated from external navigation aids. In the following, an introduction to these topics will be presented. Some simplifications will be made, especially in terms of the nature of the obstacles. The reader is encouraged to consult the list of references at the end of the chapter on mapping and localization combined, since an in-depth treatment of this topic is outside the scope of this book.

8.2 Sensors for Obstacle Detection and Geo-Registration

One important characteristic of a sensor to be used for obstacle avoidance is its operating range. Clearly, the greater the range, the more advance warning that can be provided to the robot. The speed at which the robot is moving combined with the wheels or tracks and the type surface it is traveling on determine the

Mobile Robots: Navigation, Control and Sensing, Surface Robots and AUVs,
Second Edition. Gerald Cook and Feitian Zhang.
© 2020 by The Institute of Electrical and Electronics Engineers, Inc.
Published 2020 by John Wiley & Sons, Inc.

distance required to bring the robot to a stop. The range of the obstacle detection sensor must exceed this distance, or conversely, the robot must be operated at a speed such that its stopping distance does not exceed the sensor range.

One type of sensing device is the scanning laser, which has ranges in excess of 30 m. Such devices have been used in the DARPA Grand Challenge. These lasers are capable of taking range measurements at yaw increments of a degree or less as they scan back and forth over the operating range, which may be as large as 180°. The results of this type of scan provide the information required for constructing a map of obstacles in the immediate vicinity of the robot. This map would be most easily constructed in vehicle coordinates, but could be converted to earth coordinates given robot position and heading.

Light Detection and Ranging (Lidar) is an active sensor, similar to a radar, that transmits laser pulses to a target and records the time it takes for the pulse to return to the sensor receiver. Lidar uses much shorter wavelengths of the electromagnetic spectrum, typically in the ultraviolet, visible, or near infrared range. In general, it is possible to image a feature or object only about the same size as the wavelength, or larger. A laser typically has a very narrow beam that allows the mapping of physical features with very high resolution compared with radar. Lidar has been used in adaptive cruise control systems for automobiles. Here a Lidar device is mounted on a front part of the vehicle, such as the bumper, to monitor the distance between the vehicle and any vehicle in front of it. In the event the vehicle in front slows down or is too close, the system applies the brakes to slow the vehicle. When the road ahead is clear, the system allows the vehicle to accelerate to a speed that has been preset by the driver.

Ultrasound refers to sounds or pressure waves at frequencies above those within the human hearing range. Typically the frequencies used are between 20 and 500 kHz which places them above human hearing and below AM radio. A common use of ultrasound is in range finding, which is also called sonar (sound navigation and ranging). This works similarly to radar. An ultrasonic pulse is generated in a particular direction. If there is an object in the path of this pulse, part or all of the pulse will be reflected back to the transmitter as an echo and can be detected through the receiver path. By measuring the difference in time between the pulse being transmitted and the echo being received, it is possible to determine how far away the object is. The speed of sound in the atmosphere varies with pressure and at sea level has a nominal value of 1,065 ft/s.

The higher frequency systems provide a narrower beam which can provide greater detail of the object being sensed and greater precision of its range. Systems operating at lower frequencies have a wider beam but can measure greater distances with 25 kHz sensors advertised to measure from 0.3 m up to 60 m with the aid of signal processing.

Regardless of the type of sensors used, the information provided allows one to sequentially locate the detected obstacles in the current vehicle coordinates.

Ideally one would want to then convert the obstacle locations to some common coordinate system and compile a map of all the detected obstacles. This map would be useful for future obstacle avoidance as the robot goes about its task of searching for objects of interest.

Example 1 *A mobile robot has used its sensors to geo-register two detected obstacles. The coordinates for these obstacles in local coordinates are $x_1 = 3$ and $y_1 = 10$ and $x_2 = -10$ and $y_2 = 25$. Plot these obstacles on a grid surrounding each with a 2 m radius as an approximation to the size of the obstacles.*

Solution 1

Figure 8.1 shows the locations of the obstacles and the additional boundary.

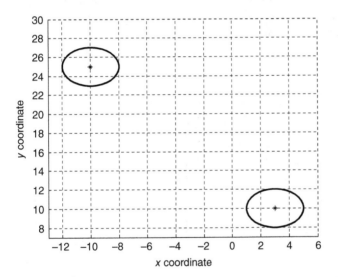

Figure 8.1 Plot showing the coordinates of the two detected obstacles.

8.3 Dead Reckoning Navigation

When the robot is isolated from external navigation aids such as GPS satellites, one could still use inertial navigation or odometry for tracking the robot in between external navigation updates. In the case of inertial navigation, the buildup of drift over time may limit its application to relative motion. Wheel slippage causes the same type of problem with odometry. Example 2 illustrates the buildup of errors possible when relying totally on dead reckoning.

Example 2 *A mobile robot using dead reckoning via odometry travels along what was intended to be a straight line. The wheels have a radius of 0.15 m. Each side underwent 30 simultaneous rotations; however, the wheel on the left side spun without moving forward at the very beginning of the trajectory while the wheel on the right side traveled forward 0.1 m. Determine the path based on dead reckoning as well as the actual path. The width of the robot is 1 m.*

Solution 2

Based on 30 simultaneous rotations of the wheels with a circumference of $2\pi R$ where R is 0.15 m, the distance traveled by each side would be 28.2743 m. The computed trajectory would be along the y axis.

Now utilizing the knowledge that the right side advanced forward 0.1 m at the beginning of the trajectory while the left side was stationary and recalling that the robot width is 1 m means that there was an undetected robot turn to the left by the amount $\psi = tan^{-1}(0.1/1)$ or $\psi = 0.0997$ rad $= 5.71°$. Thus, the actual path was a straight line headed off to the left by this computed angle. A plot of the trajectories is shown in Figure 8.2.

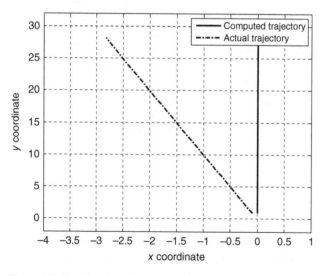

Figure 8.2 Plot showing the computed trajectory and the actual trajectory.

The above example illustrates how a small heading error at the beginning of a trajectory can propagate into a large displacement error when using dead reckoning. In the absence of better navigation, this would in turn create errors in geo-registration of any detected obstacles. In addition to the geo-registration errors caused by imperfect knowledge of the robot position, there are also errors caused by imperfect knowledge of the robot orientation. The next example illustrates this point with navigation errors in both position and heading of the robot.

Example 3 *An obstacle is detected at coordinates $x = 3$ and $y = 10$ in robot coordinates. Take the robot coordinates at the time of this detection as the origin of a local coordinate system, i.e., the robot is at $x_l = 0$ and $y_l = 0$ with heading $\psi_l = 0$ when the obstacle is detected at $x_l = 3$ and $y_l = 10$. Next, the robot moves to a location computed via imperfect navigation to be $x_l = 0$ and $y_l = 15$ with computed heading $\psi_l = 0$. At that time, another obstacle is detected and correctly located in robot coordinates at $x_r = -10$ and $y_r = 10$. In the local coordinate system its location is computed to be $x_l = -10$ and $y_l = 25$.*

a) *Plot the coordinates of the detected obstacles in the x_l, y_l plane. Surround the coordinates of the detected obstacles with a circle of radius 2 as an approximation of the space occupied by the obstacle.*

b) *Assume that the computed position and heading of the robot at the time of the second obstacle detection are erroneous and that in fact the robot's location was $x_l = 0.5$ and $y_l = 14$ with heading $\psi_l = -0.1\pi$. Compute the actual location of the second obstacle and plot its coordinates. Surround the coordinates of the detected obstacles with a circle of radius 2 as an approximation of the space occupied by the obstacle.*

Solution 3

The second obstacle was correctly located in robot coordinates at $x_r = -10$ and $y_r = 10$, but the conversion to local coordinates was erroneous because the robot heading and location were incorrect. The location of the robot in the local coordinate system was actually $x_l = 0.5$ and $y_l = 14$ with heading $\psi_l = -0.02\pi$. Therefore, the second obstacle is at
$$\begin{bmatrix} x_l \\ y_l \end{bmatrix} = \begin{bmatrix} \cos 0.02\pi & \sin 0.02\pi \\ -\sin 0.02\pi & \cos 0.02\pi \end{bmatrix} \times \begin{bmatrix} -10 \\ 10 \end{bmatrix} + \begin{bmatrix} 0.5 \\ 14 \end{bmatrix}$$
or $x_l = -8.8524$ and $y_l = 24.6082$. Figure 8.3 shows the computed coordinates of the obstacles and Figure 8.4 shows their actual coordinates.

The above example illustrates how a map of detected obstacles may be in error when the robot coordinates are determined by erroneous navigation means.

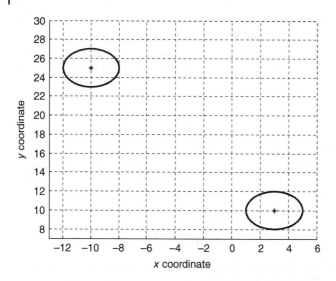

Figure 8.3 Plot showing computed coordinates of two detected obstacles.

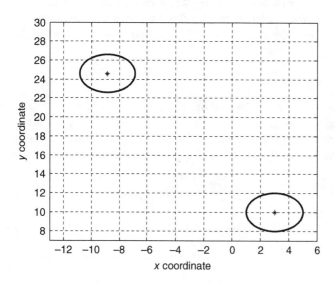

Figure 8.4 Plot showing actual coordinates of two detected obstacles.

8.4 Use of Previously Detected Obstacles for Navigation

Any planned paths for the robot must take into account the map of obstacles, and the path must be modified as necessary to avoid the obstacles. This is a primary use of the obstacle map. In addition to this application, the following example illustrates how one might also use this map of obstacles as an aid in navigation.

Example 4   *A mobile robot has measured the range from itself to two previously detected and geo-registered obstacles. The coordinates of the obstacles along with their respective ranges from the robot are $x_1 = 10$, $y_1 = 25$, $d_1 = 20$ and $x_2 = -4$, $y_2 = 30$, $d_2 = 23$. Determine the location of the robot using these obstacle locations and the distances.*

Solution 4

The solution to the problem is the set of values for x and y which simultaneously satisfy

$$(x - x_1)^2 + (y - y_1)^2 = d_1^2$$

and

$$(x - x_2)^2 + (y - y_2)^2 = d_2^2$$

By carrying out the squaring operations and then subtracting the second equation from the first equation, one obtains

$$2x(x_2 - x_1) + 2y(y_2 - y_1) = d_1^2 - d_2^2 - x_1^2 + x_2^2 - y_1^2 + y_2^2$$

Solving this for y yields

$$y = -\frac{(x_2 - x_1)}{(y_2 - y_1)}x + \frac{(d_1^2 - d_2^2 - x_1^2 + x_2^2 - y_1^2 + y_2^2)}{2(y_2 - y_1)}$$

Substituting this expression into either of the first two equations yields a quadratic equation in y whose solution is well known. Using the numbers for this example one obtains for the location of the robot $x_r = 13.73$ and $y_r = 46.65$ or $x_r = 0.44$ and $y_r = 7.43$. Figure 8.5 shows the plots of the obstacle locations and the two possible robot locations.

One must use other information such as knowledge of the approximate location to decide which solution is applicable.

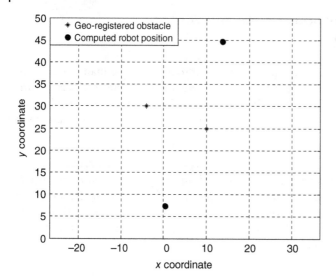

Figure 8.5 Plot showing the two geo-registered obstacles and the two computed robot locations.

The above example illustrates how one can combine distances from two geo-registered obstacles and determine the location of the mobile robot. If there are three or more obstacles, the question becomes how to incorporate the extra information. Any combination of two obstacles yields a set of solutions. The idea of a least squares solution comes to mind. A bit of reflection suggests that this is really very similar to the problem of determining the location of the robot using GPS. Here the obstacles play the role of the GPS satellites, and the dimension of the search space has been reduced from three to two.

Example 5 *A mobile robot has measured, in local coordinates, the bearing from itself to two previously detected and geo-registered obstacles. The coordinates of the obstacles along with their respective bearings from the robot are $x_1 = 10$, $y_1 = 25$, $\psi_1 = -1/8\pi$ and $x_2 = -4$, $y_2 = 30$, $\psi_2 = 1/8\pi$. Determine the location of the robot using these obstacle locations and the distances.*

Solution 5

This problem is different from the previous one in that the bearing of the ray from the robot to the obstacle is measured rather than the range. The type of measurement available is of course dependent on the sensor available. As seen in Chapter 6, one could measure bearing with a digital camera without need for a ranging device.

Letting the coordinates of the robot be represented by x and y, one may write

$$\tan \psi_1 = \frac{-(x_1-x)}{(y_1-y)}$$

and

$$\tan \psi_2 = \frac{-(x_2-x)}{(y_2-y)}$$

or

$$-(x_1-x) = (y_1-y)\tan \psi_1$$

and

$$-(x_2-x) = (y_2-y)\tan \psi_2$$

These two linear equations can be solved for x and y. Using the numbers for this example yields x = 4.0919 and y = 10.4645. Figure 8.6 shows the coordinates of the two obstacles and the computed coordinates of the robot.

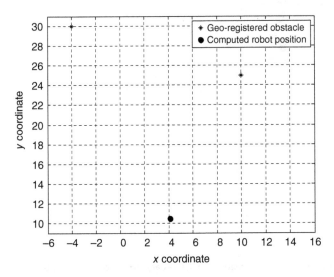

Figure 8.6 Plot showing the two geo-registered obstacles and the computed robot location using bearings only measurements.

The previous examples illustrate that if one has prior knowledge of the surroundings, sensed obstacles may provide an alternative means of navigation or a means of re-calibrating imperfect navigation to offset drift. However, if the map is imperfect, then the resulting computations will yield imperfect results. This is illustrated in the Example 6.

Example 6 *A mobile robot has measured the range from itself to two previously detected and geo-registered obstacles. The computed coordinates of the obstacles are slightly in error. These computed coordinates along with their respective ranges from the robot are $x_1 = 10.8$, $y_1 = 24.2$, $d_1 = 20$ and $x_2 = -4.5$, $y_2 = 31.2$, $d_2 = 23$. Determine the location of the robot using the computed obstacle locations. Compare the results with those of the earlier example where the true coordinates for the obstacles were known.*

Solution 6

Using the computed coordinates of the obstacles, the robot position was computed to be $x_r = 14.7358$ and $y_r = 43.8089$ or $x_r = -1.4638$ and $y_r = 8.4013$. A plot of these coordinates is shown in Figure 8.7.

The actual location based on correct coordinates for the obstacles is repeated here for comparison purposes $x_r = 13.73$ and $y_r = 46.65$ or $x_r = 0.44$ and $y_r = 7.43$.

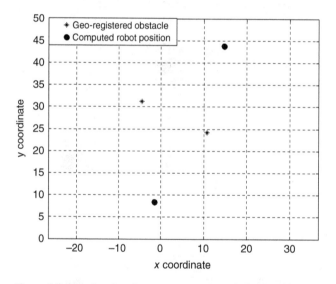

Figure 8.7 Plot showing the two geo-registered obstacles and the two computed robot locations.

The predicament should be apparent. In unfamiliar surroundings with limited navigation, one may detect and geo-register obstacles, but if the robot's position and orientation are in error, then the geo-registration will also be in error. Later, when one of these previously geo-registered obstacles is again within the field-of-view of the robot, it may be used as a source of navigation. However, if its geo-registration was in error, the aid provided toward improved robot navigation will be in error.

The navigation and mapping problems must be simultaneously addressed. When neither the robot coordinates nor the obstacle coordinates are precisely known, updates must take this into account and must weigh each piece of information according to its degree of known precision. The result of new measurements must be to not only update the position and orientation of the robot, but also to update the coordinates of the previously geo-registered obstacles.

8.5 Simultaneous Corrections of Coordinates of Detected Obstacles and of the Robot

The model for estimates in such a situation as that described at the end of the previous section must include as its state the obstacle locations as well as the location and orientation of the robot. The measurements will include terms such as the relative location of obstacles with respect to the robot. The precision of knowledge of the state of the robot and the precision of knowledge of the location of the obstacle will be reflected through the covariance matrices of each. The covariance of the measurement noise will also play a role. The impact of updates as a result of measurements will depend on these covariances and their relative magnitudes.

Consider a simple one-dimensional problem where the robot, which is stationary, and the obstacle are constrained to lie on a line. There are current estimates of the location of each. Now suppose a measurement of the distance from the robot to the obstacle is taken. If the result of this distance measurement is greater than what the estimated locations would yield, then the new estimates must be such as to have the robot and obstacle farther apart depending on the reliability of the measurement. Which estimate must move the most depends on the relative covariances of the robot location and the obstacle location.

For the more general problem, one can model the composite system as

$$\begin{bmatrix} X_{robot}(k+1) \\ X_{obstacle}(k+1) \end{bmatrix} = \begin{bmatrix} f(X_{robot}(k), u_{robot}(k)) \\ X_{obstacle}(k) \end{bmatrix} + \begin{bmatrix} w_{robot}(k) \\ w_{obstacle}(k) \end{bmatrix}$$

with output

$$y(k+1) = h(X_{robot}(k+1), X_{obstacle}(k+1)) + v(k+1)$$

One then proceeds to develop a Kalman Filter to estimate the total state of the system. Here this includes the state of the robot as well as the location of the obstacle. In the above model, it has been assumed that the obstacle is stationary. Nevertheless, the inclusion of the disturbance term in the obstacle model is required in order to provide the possibility of updates in the obstacle location as more measurements are taken. Provision has been made in the robot model for possible robot motion between measurements. The exact form of the h vector depends on what specifically is being measured. It could be the distance from the robot to the obstacle. It might also include the yaw angle of the vector from the robot to the obstacle.

This approach is illustrated in Example 7.

Example 7 *Consider the situation where a robot has previously computed its own location and that of an obstacle in a local coordinate system. The robot once more measures the x and y distances to the obstacle. Use these distance measurements to update the robot location and the obstacle location. The covariance of error in the robot location is*

$$P_{robot} = \begin{bmatrix} P_{robot-x} & 0 \\ 0 & P_{robot-y} \end{bmatrix}$$

and the covariance of error in the obstacle location is

$$P_{obstacle} = \begin{bmatrix} P_{robstacle-x} & 0 \\ 0 & P_{obstacle-y} \end{bmatrix}$$

The covariance of the distance measurements is given by $R = \begin{bmatrix} R_x & 0 \\ 0 & R_y \end{bmatrix}$.
Assume that the robot has been stationary between measurements so that $\begin{bmatrix} \hat{x}_{robot}(k+1/k) \\ \hat{y}_{robot}(k+1/k) \end{bmatrix} = \begin{bmatrix} \hat{x}_{robot}k/k \\ \hat{y}_{robot}k/k \end{bmatrix}$. *A similar equation would hold for the stationary obstacle.*

Solution 7

The estimation equation is given by $\hat{X}(k+1/k+1) = \hat{X}(k+1/k) + K(k+1)(Y(k+1) - H\hat{X}(k+1/k))$, *where the gain is given by* $K(k+1) = P(k+1/k)H^T[HP(k+1/k)H^T + R]^{-1}$.

Take the state vector to be $X = \begin{bmatrix} x_{robot} \\ y_{robot} \\ x_{obstacle} \\ y_{obstacle} \end{bmatrix}$. *The output measurements are the x*

and y distances measured in the positive direction from the robot to the obstacle.
$y_1 = x_{obstacle} - x_{robot}$ *and* $y_2 = y_{obstacle} - y_{robot}$ *yielding* $H = \begin{bmatrix} -1 & 0 & 1 & 0 \\ 0 & -1 & 0 & 1 \end{bmatrix}$.

The result of applying these definitions and operations to the estimation problem yields on an elemental basis

$$\hat{x}_{robot}(k+1/k+1) = \hat{x}_{robot}(k/k) - \frac{P_{robot-x}}{P_{robot-x} + P_{obstacle-x} + R_x}(y_1 - (\hat{x}_{obstacle}(k/k) - \hat{x}_{robot}(k/k)))$$

$$\hat{y}_{robot}(k+1/k+1) = \hat{y}_{robot}(k/k) - \frac{P_{robot-y}}{P_{robot-y} + P_{obstacle-y} + R_y}(y_2 - (\hat{y}_{obstacle}(k/k) - \hat{y}_{robot}(k/k)))$$

$$\hat{x}_{obstacle}(k+1/k+1) = \hat{x}_{obstacle}(k/k) + \frac{P_{obstacle-x}}{P_{robot-x} + P_{obstacle-x} + R_x}(y_1 - (\hat{x}_{obstacle}(k/k) - \hat{x}_{robot}(k/k)))$$

$$\hat{y}_{obstacle}(k+1/k+1) = \hat{y}_{obstacle}(k/k) + \frac{P_{obstacle-y}}{P_{robot-x} + P_{obstacle-x} + R_x}(y_2 - (\hat{y}_{obstacle}(k/k) - \hat{y}_{robot}(k/k)))$$

In order to perform some numerical comparisons let $\hat{x}_{robot}(k/k) = 10$, $\hat{y}_{robot}(k/k) = 0$, $\hat{x}_{obstacle}(k/k) = 50$, *and* $\hat{y}_{obstacle}(k/k) = 0$. *Assume that* $y_1 = 37$ *and* $y_2 = 3$.

Tables 8.1a and 8.1b give the resulting updated estimates for the specified covariances.

Note that both the robot and obstacle locations are updated. The corrections on robot and obstacle locations are equal because of the equal values for their covariances. The distances computed based on the new estimates do not match the measurements perfectly because the measurements themselves are not totally reliable.

If the covariances are not all equal, the residuals have different affects on the estimates as Tables 8.2a and 8.2b illustrate. Here ε is intended to represent a very small positive number.

In this second case, it is seen from the table on covariances that the distance measurements are very reliable and that the estimated location of the robot is

Table 8.1a First set of covariances for above example.

$P_{robot-x}$	$P_{obstacle-x}$	R_x	$P_{robot-y}$	$P_{obstacle-y}$	R_y
1.0	1.0	1.0	1.0	1.0	1.0

Table 8.1b Estimates of obstacle location and robot location before and after measurements.

	x_{robot}	y_{robot}	$x_{obstacle}$	$y_{obstacle}$
Estimate before measurement	10	0	50	0
Estimate after measurement	11	−1	49	1

Table 8.2a Second set of covariances for above example.

$P_{robot-x}$	$P_{obstacle-x}$	R_x	$P_{robot-y}$	$P_{obstacle-y}$	R_y
ε	1.0	ε	ε	1.0	ε

Table 8.2b Estimates of obstacle location and robot location before and after measurements.

	x_{robot}	y_{robot}	$x_{obstacle}$	$y_{obstacle}$
Estimate before measurement	10	0	50	0
Estimate after measurement	10	0	47	3

also very reliable, while the estimation of the obstacle location is relatively unreliable. Thus, the update results in a new estimate for the obstacle location but the same estimate for the robot location. Here the computed distances between the estimated location of the robot and the estimated location of the obstacle in both dimensions match the measurements because of their reliability.

The foregoing is intended to introduce the reader to some of the considerations one is faced with when navigation capability is limited and obstacles are present. For situations where obstacles are not isolated but rather are frequent and distributed, as would be the case inside a building, this problem becomes much more difficult. Detections of boundaries of rooms, doorways, and furnishings are all part of the problem. Whatever usable navigation is on the robot must also be utilized and incorporated as is appropriate for its degree of accuracy. Clearly this problem area is complex with errors in one part of the process feeding into the other and vice versa. This is why simultaneous localization and mapping (SLAM) is such an important area of ongoing research within the area of mobile robotics.

Exercises

1 There are two obstacles at x, y locations of −10, 20 and 18, 29 in local coordinates. Measurements taken from the mobile robot to each obstacle produce distances of 28 and 39, respectively. Find the two possible locations of the robot.

2 **A** Odometer readings for the two rear wheels of a mobile robot are as follows: $\theta_{left}(k) = k$; $k = 1, 2, ..., 200$ and $\theta_{right}(k) = k + k^2/1,000$; $k = 1,$

2, ..., 200 all in radians. The wheel radius is 0.15 m. Using dead reckoning determine and plot the robot trajectory in x, y space. Also plot robot heading versus k.

B At time $k = 200$ an obstacle is detected directly in front of the robot at a distance of 10 m. Determine the x, y coordinates of the obstacle.

3 Consider a one-dimensional location problem where the mobile robot is estimated to be at $\hat{y}_{robot} = 10$ m and an obstacle is estimated to be at $\hat{y}_{obstacle} = 30$ m. With no robot motion having occurred, a new measurement is made of the distance from the robot to the obstacle and it is found to be 23 m. Compute new estimates of both the robot location and the obstacle location for the different sets of covariance values shown in the table below. Here ε is intended to represent a very small number.

P_{robot}	1.0	1.0	ε	1.0
$P_{obstacle}$	1.0	ε	1.0	1.0
R_{meas}	ε	1.0	1.0	1.0

A mobile robot sees three obstacles whose coordinates have been previously estimated. The distances to the obstacles are measured and are to be used to estimate the location of the robot. Develop and describe a scheme for using these three pieces of information to determine the two coordinates of the robot location. Hint: Read the discussion following the example where distances to two obstacles are used to determine the location of the robot.

References

Durrant-Whyte, H. and Bailey, T., "Simultaneous Localisation and Mapping (SLAM): Part I The Essential Algorithms," *Robotics and Automation Magazine*, Vol. **13** (2006), pp. 99–110.

Kansal, S., Cook, G., Sherbondy, K., and Amazeen, C., "Use of Fiducials and Un-surveyed Landmarks as Geo-Location Tools in Vehicular Based Landmine Search," *IEEE Transactions on Geoscience and Remote Sensing*, Vol. **43**, No. 6 (June 2005), pp. 1432–1439.

Leonard, J. J. and Durrant-Whyte, H. F., "Simultaneous Map Building and Localization for an Autonomous Mobile Robot," *Proceedings IROS '91. IEEE/RSJ International Workshop on Intelligent Robots and Systems '91. Intelligence for Mechanical Systems* (pp. 1442–1447, IEEE, Osaka, Japan, November 3, 1991).

Smith, R. C., Self, M., and Cheeseman, P., "Estimating Uncertain Spatial Relationships in Robotics," *Proceedings of the Second Annual Conference on*

Uncertainty in Artificial Intelligence. UAI '86, University of Pennsylvania, Philadelphia, PA, Elsevier (pp. 435–461, 1986).

Wang, Chieh-Chih and Thorpe, C., "Simultaneous Localization and Mapping with Detection and Tracking of Moving Objects," *Robotics and Automation,* Vol. **3** (2002), pp. 2918–2924.

Wolf, Denis and Sukhatme, Gaurav S., "Online Simultaneous Localization and Mapping in Dynamic Environments," *Proceedings of the International Conference on Robotics and Automation ICRA* (IEEE, New Orleans, LA, April 26, 2004).

9

Operating a Robotic Manipulator

9.1 Introduction

This chapter is devoted to the treatment of a simple three-degree-of-freedom robotic manipulator. Several possible tasks for mobile robots were discussed in the beginning of this book. Some of these, such as retrieval of a target, material transfer, and inspection of a surface could require the use of a robotic manipulator mounted on the mobile robot. Another application would be for positioning a sensor close to a target for a more detailed analysis. The forward and inverse kinematic equations are presented and analyzed.

9.2 Forward Kinematic Equations

Figure 9.1 shows the manipulator to be studied.

The workspace for this robotic manipulator is a portion of a sphere of radius $l_2 + l_3$ centered l_1 units above the base of the robotic manipulator. The lower limits of the workspace will depend on practical considerations in the construction of the manipulator.

In analyzing this manipulator, first we shall write the equations for expressing the end-effector position in Cartesian coordinates in terms of the robotic joint angles. These are called the forward kinematic equations. A frame attached to the base of the robotic manipulator shall be the coordinate frame used for this specification. Clearly a point specified in another coordinate system such as vehicle coordinates or earth coordinates could be converted to these coordinates and vice versa.

First note that the horizontal component of the vector from the base of the robotic manipulator to the end-effector is given by

$$\sqrt{x^2 + y^2} = l_2 \cos \theta_2 + l_3 \cos (\theta_2 + \theta_3) \tag{9.1}$$

Mobile Robots: Navigation, Control and Sensing, Surface Robots and AUVs,
Second Edition. Gerald Cook and Feitian Zhang.
© 2020 by The Institute of Electrical and Electronics Engineers, Inc.
Published 2020 by John Wiley & Sons, Inc.

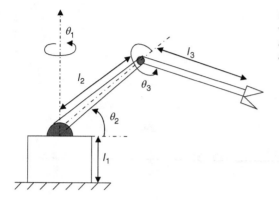

Figure 9.1 Robotic manipulator: waist, shoulder, forearm with angles θ_1, θ_2, and θ_3 and links 1, 2, and 3.

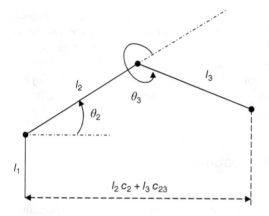

Figure 9.2 Schematic illustration of the robotic manipulator: side view showing links 1, 2, and 3.

This can be seen from Figure 9.2 in the plane of the links.

This may be further broken down into its two components. Figure 9.3 shows the top view of the manipulator.

The x component is given by this horizontal component combined with the waist angle and is given by

$$x = -\{l_2 \cos \theta_2 + l_3 \cos (\theta_2 + \theta_3)\} \sin \theta_1 \tag{9.2}$$

Likewise the y component is given by

$$y = \{l_2 \cos \theta_2 + l_3 \cos (\theta_2 + \theta_3)\} \cos \theta_1 \tag{9.3}$$

Finally, the vertical projection z of the end-effector from the base is given by

$$z = l_1 + l_2 \sin \theta_2 + l_3 \sin (\theta_2 + \theta_3) \tag{9.4}$$

These expressions for x, y, and z define the forward kinematic equations for the manipulator, i.e., they define the Cartesian coordinates of the end-effector in

Figure 9.3 Schematic illustration of the robotic manipulator: top view showing links 2 and 3.

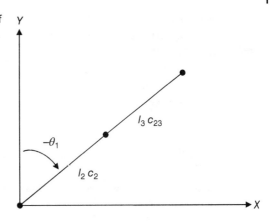

terms of the robot joint angles. For more complex robotic manipulator configurations one may utilize the Denavit–Hartenberg parameters to describe each link and from these form the homogeneous transformation matrix which describes end-effector position and orientation as a function of joint angles.

Example 1   *Take $\theta_1 = 0°$, $\theta_2 = 0°$, and $\theta_3 = 0°$. Let $L1 = 1$, $L2 = 2$, and $L3 = 2$. Use the forward kinematic equations to determine the position of the end-effector.*

Solution 1

$x = 0$, $y = 4$, $z = 1$.

Example 2   *Take $\theta_1 = 0°$, $\theta_2 = 90°$, and $\theta_3 = -90°$. Let $L1 = 1$, $L2 = 2$, and $L3 = 2$. Use the forward kinematic equations to determine the position of the end-effector.*

Solution 2

$x = 0$, $y = 2$, $z = 3$.

Example 3   *Take $\theta_1 = 0°$, $\theta_2 = 45°$, and $\theta_3 = -90°$. Let $L1 = 1$, $L2 = 2$, and $L3 = 2$. Use the forward kinematic equations to determine the position of the end-effector.*

Solution 3

$x = 0$, $y = 2.828$, $z = 1$.

Example 4   *Take $\theta_1 = 45°$, $\theta_2 = 45°$, and $\theta_3 = -90°$. Let L1 = 1, L2 = 2, and L3 = 2. Use the forward kinematic equations to determine the position of the end-effector.*

Solution 4

$x = -2$, $y = 2$, $z = 1$.

9.3   Path Specification in Joint Space

Given a trajectory, i.e., a sequence of points in joint space, one could use these equations repeatedly to determine the trajectory, or sequence of points, of the end-effector in Cartesian space.

Example 5   *Starting from reaching toward the front and swinging around to the right and up. Let $\theta_1 = -(k/20) * \pi/2$, $\theta_2 = \pi/4 + k/100$, $\theta_3 = -\pi/4$ for k = 1, 2, 3, ..., 20. Plot the path of the end-effector.*

Solution 5

The plots showing the solution to this example are shown in Figure 9.4a–d.

Example 6   *Reaching to the front and downward. Let $\theta_1(k) = 0$, $\theta_2(k) = \pi/4 - k/100$, and $\theta_3(k) = -\pi/4 + k/100$; k = 1, 2, 3, ..., 20. Plot the path of the end-effector.*

Solution 6

The plots showing the solution to this example are shown in Figure 9.5a–c.

9.4   Inverse Kinematic Equations

Often one needs to solve the inverse kinematic equations and obtain expressions for the robot joint angles in terms of the desired end-effector position in Cartesian coordinates. These solutions would permit one to start with a desired position for the end-effector and determine the joint angles required to accomplish this position. Figure 9.6 presents a schematic drawing of the side view of robotic manipulator with intermediate angles defined.

One must ensure that the specified end-effector position is within reach in order to guarantee a solution. This requires that

$$\sqrt{x^2 + y^2 + (z - l_1)^2} \leq l_2 + l_3$$

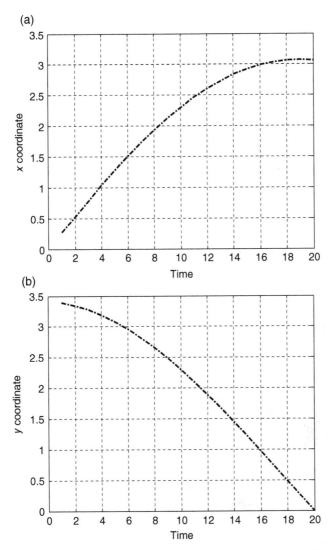

Figure 9.4 (a) Plot of x coordinate of end-effector versus time. (b) Plot of y coordinate of end-effector versus time. (c) Plot of z coordinate of end-effector versus time. (d) Path of end-effector in *xyz* space.

Given that the specified end-effector position is within reach, one may proceed. From the expressions for x and y one obtains the equation

$$\tan(\theta_1) = \frac{-x}{y}$$

(c)

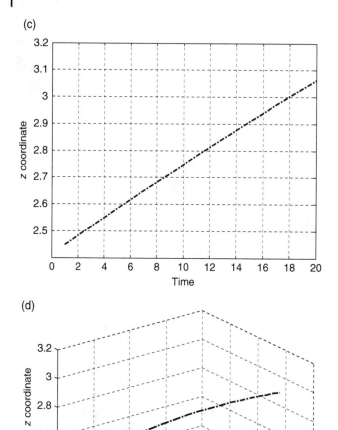

(d)

Figure 9.4 (Continued)

or

$$\theta_1 = \tan^{-1}\left(\frac{-x}{y}\right) \tag{9.5}$$

Using the law of cosines, the square of the distance from the joint 2 of the manipulator to the end-effector can be determined as

$$x^2 + y^2 + (z - l_1)^2 = l_2^2 + l_3^2 - 2l_2l_3 \cos(\theta_3 + \pi)$$

or

$$x^2 + y^2 + (z - l_1)^2 = l_2^2 + l_3^2 + 2l_2l_3 \cos(\theta_3)$$

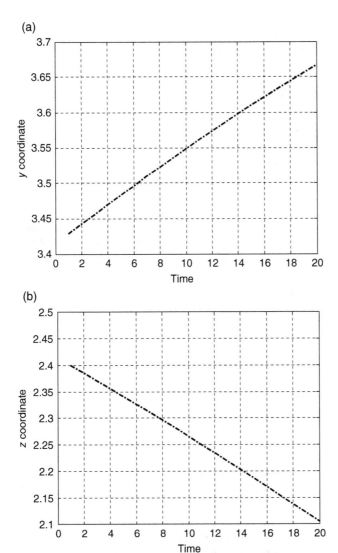

(a)

(b)

Figure 9.5 (a) Plot of *y* coordinate of end-effector versus time. (b) Plot of *z* coordinate of end-effector versus time. (c) Path of end-effector in *xyz* space.

(c)

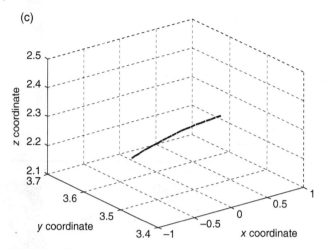

Figure 9.5 (Continued)

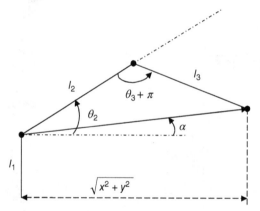

Figure 9.6 Schematic drawing of side view of robotic manipulator with intermediate angles defined.

which yields

$$\theta_3 = \cos^{-1}\left(\frac{x^2 + y^2 + (z - l_1)^2 - l_2^2 - l_3^2}{2l_2l_3}\right) \tag{9.6}$$

Note that the inverse cosine operation yields a positive or negative angle. For practical purposes, θ_3 is given the negative solution, i.e., elbow up. Thus the inverse solution is required to be between $0°$ and $-180°$. Thus if the cosine is a positive value, the angle is taken to be between $0°$ and $-90°$, and if the cosine is a negative value, the angle is taken to be between $-90°$ and $-180°$.

Consider again the side view of links 2 and 3. The angle α represents the angle of the vector from joint 2 to the end-effector with respect to horizontal. It satisfies the equation

$$\tan(\alpha) = \frac{z - l_1}{\sqrt{x^2 + y^2}}$$

or

$$\alpha = \tan^{-1}\left\{ \frac{z - l_1}{\sqrt{x^2 + y^2}} \right\} \tag{9.7}$$

Once this has been computed, one can again use the law of cosines to write

$$l_3^2 = l_2^2 + x^2 + y^2 + (z - l_1)^2 - 2l_2\sqrt{x^2 + y^2 + (z - l_1)^2}\cos(\theta_2 - \alpha)$$

which leads to

$$2l_2\sqrt{x^2 + y^2 + (z - l_1)^2}\cos(\theta_3 - \alpha) = l_2^2 - l_3^2 + x^2 + y^2 + (z - l_1)^2$$

or

$$\theta_3 = \cos^{-1}\left\{ x^2 + y^2 + (z - l_1)^2 + l_2^2 - l_3^2 \right\} \Big/ \left\{ 2l_2\sqrt{x^2 + y^2 + (z - l_1)^2} \right\} + \alpha \tag{9.8}$$

Here the solution to the inverse cosine operation is taken to be between $0°$ and $180°$, i.e., a positive cosine value corresponds to an inverse cosine between $0°$ and $90°$ and a negative cosine value corresponds to an inverse cosine between $90°$ and $180°$. The three equations, (9.5), (9.6), and (9.8), constitute the required expressions for obtaining the robotic manipulator angles given the desired Cartesian coordinates for the end-effector.

Example 7   *Let the links have length given by L1 = 1, L2 = 2, and L3 = 2. Let the coordinates of the end-effector be x = 0, y = 4, and z = 1. Find the required joint angles.*

Solution 7

$\theta_1 = 0°, \theta_2 = 0°, \text{ and } \theta_3 = 0°.$

Example 8   *Let the links have length given by L1 = 1, L2 = 2, and L3 = 2. Let the coordinates of the end-effector be x = 2, y = 2, and z = 1. Find the required joint angles.*

Solution 8

$\theta_1 = -45°, \theta_2 = 45°, and \theta_3 = -90°.$

Example 9   *Let the links have length given by L1 = 1, L2 = 2, and L3 = 2. Let the coordinates of the end-effector be x = 2, y = 2, and z = 2. Find the required joint angles.*

Solution 9

$\theta_1 = -45°, \theta_2 = 60.9°, and \theta_3 = -82.8°.$

9.5   Path Specification in Cartesian Space

In some cases, one may simply require that the end-effector of the robotic manipulator be placed at a specified location. Here one can compute the required joint angles using the inverse kinematic equations as in the examples above and then direct the controllers to drive each joint to the proper angle without regard for any coordination among the joints. In other cases, there may be a path for the motion, specified as a series of points in Cartesian coordinates. In such cases, one can solve the inverse kinematic equations for each of the points and construct the corresponding path in robot joint coordinates. The motion in joint coordinates would then be along this path. Several examples are given below.

Example 10   *Reaching out to the front to hold a sensor over an object. Let x = 0, y = (k/20) ∗ 3, and z = 1 for k = 1, 2, 3, …, 20. The links have lengths L1 = 1, L2 = 2, and L3 = 2. Determine the required joint angles.*

Solution 10

Plots illustrating the solution are shown in Figure 9.7a–e.

Example 11   *Lifting an object located in front of the robot, with link lengths L1 = 1, L2 = 2, and L3 = 2. Let x(k) = 0, y(k) = 3, and z(k) = 2 ∗ (k/20) for k = 1, 2, …, 20. Determine the required joint angles.*

Solution 11

Plots illustrating the solution are shown in Figure 9.8a–e.

Example 12   *Reaching over an obstacle to retrieve an object located at the front right of the robot. With link lengths L1 = 1, L2 = 2, and L3 = 2 let x(k) = 2 ∗ (k/20), y(k) = 2 ∗ (k/20), and z(k) = 1 + 2 ∗ sin(π ∗ k/20) for k = 1, 2, …, 20. Determine the required joint angles.*

Solution 12

Plots illustrating the solution are shown in Figure 9.9a–d.

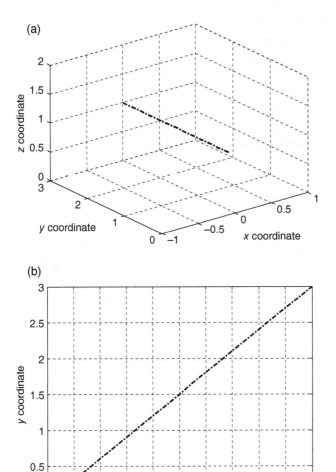

Figure 9.7 (a) Specified path of end-effector in *xyz* space. (b) *y* coordinate of end-effector versus time, $x = 0$; $z = L1 = 1$. (c) Joint angle θ_1 versus time. (d) Joint angle θ_2 versus time. (e) Joint angle θ_3 versus time.

9.6 Velocity Relationships

It is interesting to consider the velocities required by the robotic joints as a function of the velocities in Cartesian coordinates that result from the path specification. For this analysis use is made of the Jacobian. Recalling the equations for

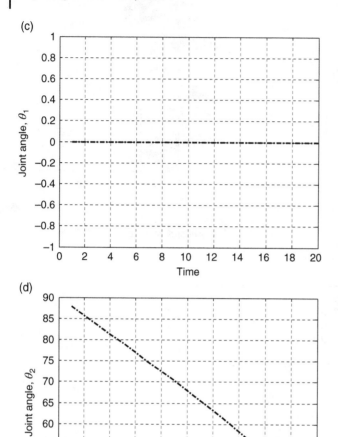

Figure 9.7 (Continued)

x, y, and z in terms of the joint angles, one can take partial derivatives of the Cartesian variables with respect to the joint angles and obtain

$$\begin{bmatrix} \dot{x} \\ \dot{y} \\ \dot{z} \end{bmatrix} = J \begin{bmatrix} \dot{\theta}_1 \\ \dot{\theta}_2 \\ \dot{\theta}_3 \end{bmatrix}$$

(9.8a)

(e)

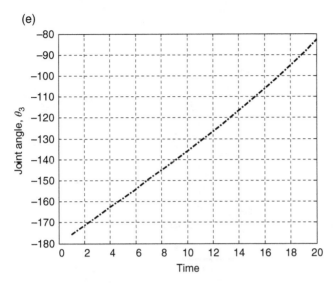

Figure 9.7 (Continued)

or

$$\begin{bmatrix} \dot{x} \\ \dot{y} \\ \dot{z} \end{bmatrix} = \begin{bmatrix} -l_2 c_1 c_2 - l_3 c_1 c_{23} & l_2 s_1 s_2 + l_3 s_1 s_{23} & l_3 s_1 s_{23} \\ -l_2 s_1 c_2 - l_3 s_1 c_{23} & -l_2 c_1 s_2 - l_3 c_1 s_{23} & -l_3 c_1 s_{23} \\ 0 & l_2 c_2 + l_3 c_{23} & l_3 c_{23} \end{bmatrix} \begin{bmatrix} \dot{\theta}_1 \\ \dot{\theta}_2 \\ \dot{\theta}_3 \end{bmatrix} \qquad (9.8b)$$

This equation shows how the joint velocities affect the velocity of the end-effector in Cartesian space. By formally taking the inverse of the Jacobian matrix we obtain

$$\begin{bmatrix} \dot{\theta}_1 \\ \dot{\theta}_2 \\ \dot{\theta}_3 \end{bmatrix} = \begin{bmatrix} -l_2 c_1 c_2 - l_3 c_1 c_{23} & l_2 s_1 s_2 + l_3 s_1 s_{23} & l_3 s_1 s_{23} \\ -l_2 s_1 c_2 - l_3 s_1 c_{23} & -l_2 c_1 s_2 - l_3 c_1 s_{23} & -l_3 c_1 s_{23} \\ 0 & l_2 c_2 + l_3 c_{23} & l_3 c_{23} \end{bmatrix}^{-1} \begin{bmatrix} \dot{x} \\ \dot{y} \\ \dot{z} \end{bmatrix} \qquad (9.9)$$

This equation shows how the velocity of the end-effector in Cartesian space affects the velocities of the robotic joints. Of particular interest are situations where J is nearly singular causing J^{-1} to be very large. For such configurations extremely large joint velocities may be required to achieve even moderate

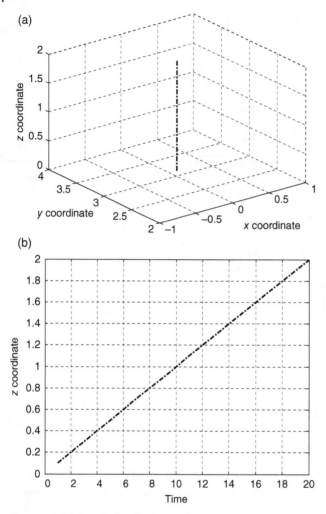

(a)

(b)

Figure 9.8 (a) Specified path of end-effector in *xyz* space. (b) *z* coordinate of end-effector versus time. (c) Joint angle θ_1 versus time. (d) Joint angle θ_2 versus time. (e) Joint angle θ_3 versus time.

Cartesian velocities. By taking the determinant of the Jacobian, after some algebra it can be seen that the Jacobian is singular whenever

$$l_2 l_3 (l_2 c_2 + l_3 c_{23})(-s_3) = 0 \tag{9.10}$$

When

$$(l_2 c_2 + l_3 c_{23}) = 0$$

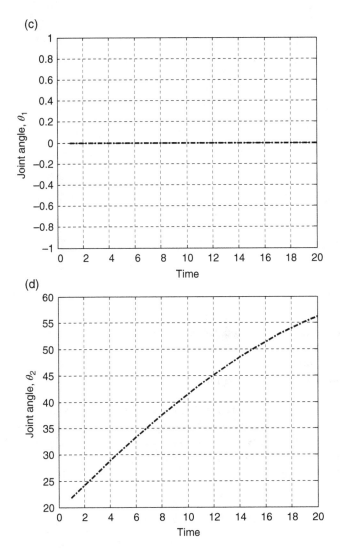

Figure 9.8 (Continued)

the end-effector is located along the axis of joint 1 and there is no possible instantaneous motion of the end-effector in the direction perpendicular to the plane of links 2 and 3. See Figure 9.10.

When $s_3 = 0$, links 2 and 3 are aligned and there is no possible motion radially. See Figure 9.11. One should avoid specifying paths which cause the robotic manipulator to even come close to these singular configurations.

(e)

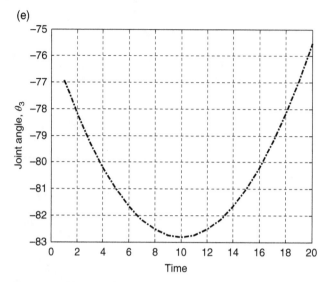

Figure 9.8 (Continued)

In the above, we see that θ_1 changed rapidly from $0°$ to $-180°$ as the robot passed near the singular configuration. The function approaches a step function as the y coordinate becomes smaller and smaller. The derivative, i.e., the velocity of the joint angle thus approaches an impulse.

Example 13 *The end-effector is moved at a constant height from a point in front of the robot to a point behind the robot passing close to a singular configuration. Let the link lengths be L1 = 1, L2 = 2, and L3 = 2. The specified path is $x = -0.01$, $y = 1 - (k/2{,}000) * 2$, and $z = 2$; for $k = 1, 2, ..., 2{,}000$. Determine the required joint angles.*

Solution 13

Plots illustrating the solution are shown in Figure 9.12a–d.

Example 14 *The end-effector is moved along a path described $x(k) = 0$, $y(k) = 2 * (1 - [k/10])$, and $z(k) = 1$ for $k = 1:9$ and $x(k) = 0$, $y(k) = 0$, and $z(k) = 1 + 2 * ([k-9]/10)$ for $k = 10:20$. This path causes the configuration to become nearly singular. Determine the required joint angles.*

Solution 14

Plots illustrating the solution are shown in Figure 9.13a–d.

In this example, it is seen that θ_2 changes rapidly with time when the end-effector is near the singular point, $x = 0$, $y = 0$, $z = 1$.

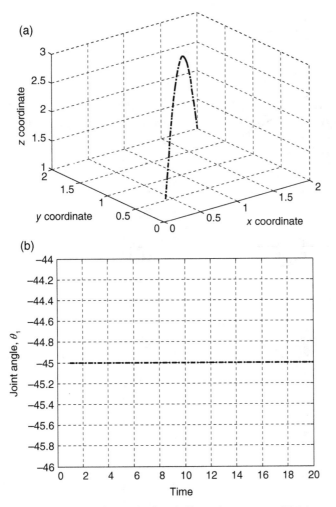

Figure 9.9 (a) Specified path of end-effector in *xyz* space. (b) Joint angle θ_1 versus time. (c) Joint angle θ_2 versus time. (d) Joint angle θ_3 versus time.

9.7 Forces and Torques

The Jacobian was used to develop the relationship between the velocities of the end-effector in Cartesian space and the angular velocities in joint space.

$$\begin{bmatrix} \dot{x} \\ \dot{y} \\ \dot{z} \end{bmatrix} = J \begin{bmatrix} \dot{\theta}_1 \\ \dot{\theta}_2 \\ \dot{\theta}_3 \end{bmatrix} \tag{9.11}$$

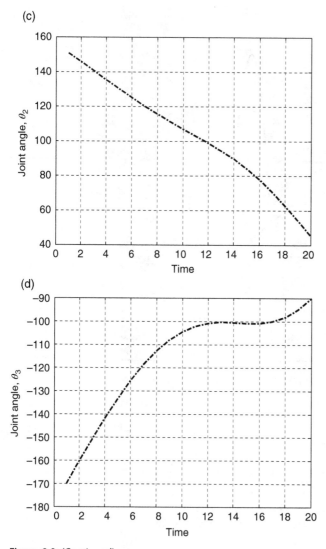

(c)

(d)

Figure 9.9 (Continued)

This matrix can also be used to develop the relationship between joint torques and end-effector forces. Now the total power delivered at the end-effector must be equal to the total power delivered by the joint actuators, i.e.,

$$\begin{bmatrix} F_x \\ F_y \\ F_z \end{bmatrix}^T \begin{bmatrix} \dot{x} \\ \dot{y} \\ \dot{z} \end{bmatrix} = \begin{bmatrix} \tau_1 \\ \tau_2 \\ \tau_3 \end{bmatrix}^T \begin{bmatrix} \dot{\theta}_1 \\ \dot{\theta}_2 \\ \dot{\theta}_3 \end{bmatrix} \qquad (9.12a)$$

Figure 9.10 One of the singular configurations for the robotic manipulator.

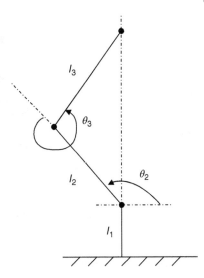

Figure 9.11 Other singular configurations for the robotic manipulator.

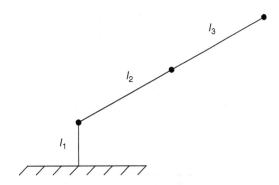

or by using the relationships between velocities

$$
\begin{bmatrix} F_x \\ F_y \\ F_z \end{bmatrix}^T J \begin{bmatrix} \dot{\theta}_1 \\ \dot{\theta}_2 \\ \dot{\theta}_3 \end{bmatrix} = \begin{bmatrix} \tau_1 \\ \tau_2 \\ \tau_3 \end{bmatrix}^T \begin{bmatrix} \dot{\theta}_1 \\ \dot{\theta}_2 \\ \dot{\theta}_3 \end{bmatrix}
\tag{9.12b}
$$

For the above to be true,

$$
\begin{bmatrix} \tau_1 \\ \tau_2 \\ \tau_3 \end{bmatrix}^T = \begin{bmatrix} F_x \\ F_y \\ F_z \end{bmatrix}^T J
\tag{9.13a}
$$

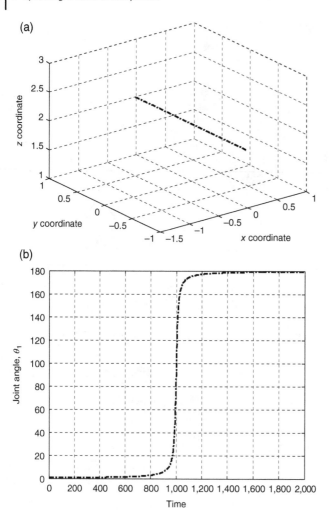

Figure 9.12 (a) Specified of end-effector in *xyz* space. (b) Joint angle θ_1 versus time. (c) Joint angle θ_2 versus time. (d) Joint angle θ_3 versus time.

or

$$\begin{bmatrix} F_x \\ F_y \\ F_z \end{bmatrix} = \left[J^T \right]^{-1} \begin{bmatrix} \tau_1 \\ \tau_2 \\ \tau_3 \end{bmatrix} \qquad (9.13b)$$

Since the determinant of a matrix is the same as the determinant of its transpose, the very same configurations that caused the Jacobian to be nearly singular

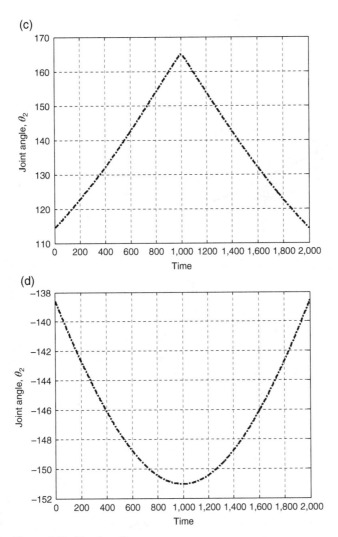

Figure 9.12 (Continued)

and required very large joint velocities for given end-effector velocities provide very large forces at the end-effector for given torques at the joints.

For the situation where $\sin(\theta_3) = 0$, the robotic manipulator has great stiffness in the radial direction along links 2 and 3. For the other singular condition where $l_2 \cos(\theta_2) + l_3 \cos(\theta_2 + \theta_3) = 0$, the robotic manipulator has great stiffness in the direction normal to the plane containing links 2 and 3.

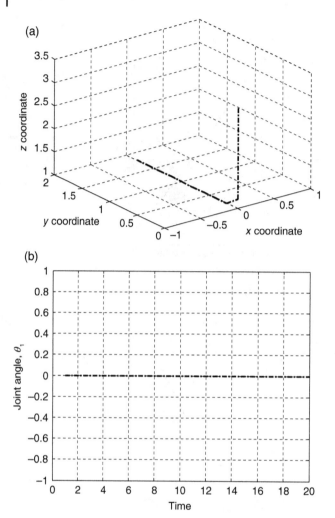

Figure 9.13 (a) Specified path of end-effector in *xyz* space. (b) Joint angle θ_1 versus time. (c) Joint angle θ_2 versus time. (d) Joint angle θ_3 versus time.

The relationships just developed can be used to determine the torques required to accomplish a particular task where the force required at the end-effector is known. This could apply for example to the task of lifting an object of known weight.

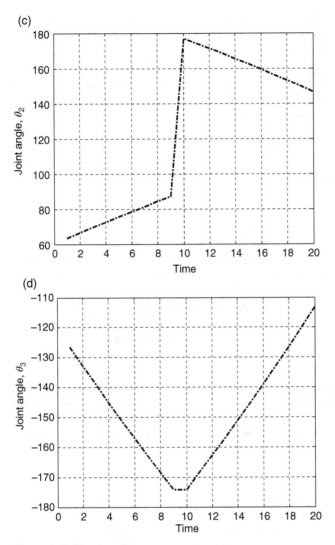

(c)

(d)

Figure 9.13 (Continued)

Exercises

1   A robotic manipulator has its joints set at $\theta_1 = \pi/3$, $\theta_2 = \pi/4$, and $\theta_3 = -\pi/2$. The robot links are of length $L1 = 1$ m, $L2 = 3$ m, and $L3 = 3$ m. Find the location of the end-effector.

2 A robotic manipulator has its joints set at $\theta_1 = -\pi/2$, $\theta_2 = -\pi/3$, and $\theta_3 = -\pi/3$. The robot links are of length $L1 = 1$ m, $L2 = 3$ m, and $L3 = 3$ m. Find the location of the end-effector.

3 A robotic manipulator is to place its end-effector at the point $x = -1$, $y = 2$, and $z = 1$. The robot links are of length $L1 = 1$ m, $L2 = 3$ m, and $L3 = 3$ m. Determine the required joint angles.

4 A robotic manipulator is to place its end-effector at the point $x = 2$, $y = 1$, and $z = 2$. The robot links are of length $L1 = 1$ m, $L2 = 3$ m, and $L3 = 3$ m. Determine the required joint angles.

5 A robotic manipulator is to be used to place an instrument near an object of interest. In terms of the coordinates of the base of the robot, the object is at $x = 2$, $y = 3$, and $z = -1$. The robot links are of length $L1 = 1$ m, $L2 = 3$ m, and $L3 = 3$ m. Find the required angles for this manipulator. If there is an ambiguity in joint 3, put the elbow up, i.e., choose the negative value for theta 3.

6 A robotic manipulator is to move the end-effector along a path specified by $x(k) = 1 - 2k/20$, $y(k) = 0.2$, and $z(k) = -1$ for $k = 1, 2, 3, ..., 20$. The robot links are of length $L1 = 1$ m, $L2 = 3$ m, and $L3 = 3$ m. Determine profiles for each of the required joint angles and plot them versus k.

7 A robotic manipulator has its joints set at $\theta_1 = -\pi/2$, $\theta_2 = 3\pi/8$, and $\theta_3 = -\pi/6$. The robot links are of length $L1 = 1$ m, $L2 = 3$ m, and $L3 = 3$ m. First find the location of the end-effector. Next find the required joint angle rates for the end-effector to be moving at an instantaneous velocity of $\dot{x} = 0$, $\dot{y} = 0$, and $\dot{z} = 2$.

References

Asada, H. and Slotine, J.-J. E., *Robot Analysis and Control*, Wiley, New York, 1986.

Craig, J. J., *Introduction to Robotics: Mechanics and Control*, Addison and Wesley, Reading, MA, 1989.

Paul, R. P., *Robot Manipulators: "Mathematics, Programming and Control,"* MIT Press, Cambridge, MA, 1982.

Sciavicco, Lorenzo and Siciliano, Bruno, *Modelling and Control of Robot Manipulators*, McGraw Hill, 1996.

Yoshikawa, Tsuneo, "Analysis and Control of Robot Manipulators with Redundancy," *Robotics Research: The First International Symposium*, MIT Press, Cambridge, MA, 1984, pp. 735–748.

Yoshikawa, Tsuneo, *Foundations of Robotics: Analysis and Control*, MIT Press, 1990.

10

Remote Sensing via UAVs

10.1 Introduction

Mobile robots are not restricted to ground vehicles. There are unmanned water vehicles, unmanned air vehicles, as well as unmanned space vehicles. In this chapter, we discuss the unmanned air vehicle when used in remote sensing. Requirements on sensor resolution and precision of vehicle attitude and position are treated.

The aircraft might use radar and/or IR as well as other sensors. It is required that objects of interest be recognizable, i.e., there need to be enough pixels on the target to permit recognition. It is also required that one be able to determine the ground coordinates of the object within some level of precision. This would enable retrieval or neutralization of the object or whatever other action might be desired.

10.2 Mounting of Sensors

In some cases, the sensor may be mounted directly to the aerial vehicle, in which case it experiences the same attitude changes as the vehicle. This could yield a somewhat jittery sensor footprint or field of view. Another approach is to mount a gimbaled sensor platform on the aerial vehicle with the gimbals having the ability to compensate for vehicle attitude changes. This would permit a steadier sensor field of view in the presence of attitude disturbances. In either case, it is necessary to very precisely know the attitude of the sensors. This can be done via attitude instrumentation for the vehicle coupled with knowledge of the motion of the gimbaled platform with respect to the vehicle, or the attitude instruments could be attached directly to the gimbaled platform.

Mobile Robots: Navigation, Control and Sensing, Surface Robots and AUVs,
Second Edition. Gerald Cook and Feitian Zhang.
© 2020 by The Institute of Electrical and Electronics Engineers, Inc.
Published 2020 by John Wiley & Sons, Inc.

10.3 Resolution of Sensors

An airborne vehicle is shown in Figure 10.1 along with a vehicle-based coordinate system. As before, y is along the longitudinal axis of the vehicle, x is to the right, and z is up. A target is shown on the ground with its location given in vehicle coordinates. It is desired to recognize the object of interest and to geo-register it. For recognition, one might assume that 25 pixels on the target would be sufficient, e.g., 5 × 5 pixels. If the target is of size 0.2 m × 0.2 m, then the pixel spacing on the ground would have to be no greater than 0.04 m or 4 cm.

A typical IR camera might have a field of view approximately equal to 12° × 16°. At an altitude of 300 m, this field of view on the ground would be approximately 63 m × 84 m. Dividing these numbers by the required resolution of 0.04 m yields 1,575 × 2,100 as the required dimension of the camera focal array in pixels. This is not unreasonable.

10.4 Precision of Vehicle Instrumentation

To examine the precision of the geo-registration task, the effects of vehicle attitude and position will be considered. The uncertainty in the x coordinate will be given by

$$\sigma^2_{x-target} = h^2\sigma^2_\phi + y^2\sigma^2_\psi + \sigma^2_{x-sensor} + \sigma^2_{x-vehicle}$$

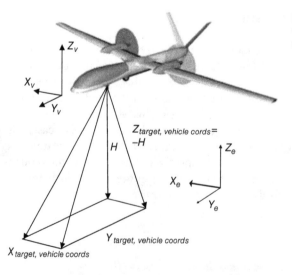

Figure 10.1 Airborne vehicle with downward-looking sensor.

The uncertainty in the y coordinate will be given by

$$\sigma^2_{y-target} = h^2 \sigma^2_\theta + x^2 \sigma^2_\psi + \sigma^2_{y-sensor} + \sigma^2_{y-vehicle}$$

An advertisement for the Ashtech ADU3 GPS array with 1 m separation between antenna elements lists accuracy in pitch and roll to be 0.8° rms and accuracy in yaw to be 0.4° rms. In terms of radians, these figures correspond to

$$\sigma^2_\psi = 4.9 \times 10^{-5}$$
$$\sigma^2_\theta = 1.95 \times 10^{-4}$$

and

$$\sigma^2_\phi = 1.95 \times 10^{-4}$$

Regarding vehicle position, this same specification sheet states accuracy of 40 cm circular error probable (CEP) for the differential mode and 3 m CEP for the autonomous mode. CEP is the radius of a circle for which the probability of the event being inside is 0.5. It differs from the elliptical confidence regions unless the standard deviations are equal in both directions. It is also more difficult to calculate than the ellipses when the standard deviations are unequal. For the case where the standard deviations are equal in both directions, the radius becomes

$$r = 1.386 \times \sigma$$

or conversely

$$\sigma = 0.7215r$$

For the CEP's stated these correspond to

$$\sigma^2_x = \sigma^2_y = (0.72 \times 0.40)^2 = 8.3 \times 10^{-2}$$

when operating in the differential mode (DGPS) and

$$\sigma^2_x = \sigma^2_y = (0.72 \times 3)^2 = 4.7$$

when operating in the autonomous mode. Clearly GPS in the autonomous mode does not yield a high degree of location precision.

10.5 Overall Geo-Registration Precision

For the sensor, the precision is limited by the quantization caused by the digital nature of the camera. Take Δ as the pixel size on the ground. Assuming that errors in localizing the source of a signal occurring within a given pixel range from $-\Delta/2$ to $\Delta/2$ with a uniform distribution, one is able to determine the corresponding variance to be

$$\sigma^2 = \Delta^2/12$$

For

$$\Delta = 0.04 \text{ m}$$

this yields

$$\sigma^2_{x-sensor} = \sigma^2_{y-sensor} = 1.33 \times 10^{-4} \text{m}^2$$

Now using the equations above with the numbers just obtained yields

$$\sigma^2_{x-target} = h^2 \times 1.95 \times 10^{-4} + y^2 \times 4.95 \times 10^{-5} + 1.33 \times 10^{-4} + 8.3 \times 10^{-2}$$

$$\sigma^2_{y-target} = h^2 \times 1.95 \times 10^{-4} + x^2 \times 4.9 \times 10^{-5} + 1.33 \times 10^{-4} + 8.3 \times 10^{-2}$$

It is clear that if the sensor meets the resolution requirements for target recognition, it will not contribute appreciably to the geo-registration error. Also, if DGPS can be used, the vehicle position error will not add too much to the error. Now taking as a typical target location with respect to the vehicle,

$$x = 15 \text{ m}$$

$$y = 20 \text{ m}$$

$$z = -300 \text{ m}$$

$$\sigma^2_{x-target} = 17.55 + 0.02 + 1.33 \times 10^{-4} + 8.3 \times 10^{-2}$$

$$\sigma^2_{y-target} = 17.55 + 0.011 + 1.33 \times 10^{-4} + 8.3 \times 10^{-2}$$

The greatest source of error is the uncertainty in the vehicle pitch and roll which get magnified by the altitude of the vehicle squared. In order to improve the geo-registration accuracy to the centimeter range, the variances in pitch and roll need to be reduced by almost 100.

Multiple looks will assist in reducing the error covariance below what it would be for a single look. If the errors from look to look are independent, the variance reduces as one over the number of looks. However, this alone would not take care of the problem here. The LEICA DMC Inclinometer advertises variances of

$$\sigma^2_\psi = 4 \times 10^{-6}$$

$$\sigma^2_\theta = 5 \times 10^{-6}$$

and

$$\sigma^2_\phi = 5 \times 10^{-6}$$

Using these numbers, the calculations of geo-registration uncertainty become

$$\sigma^2_{x-target} = 0.45 + 0.0016 + 1.33 \times 10^{-4} + 8.3 \times 10^{-2} \quad \text{and}$$

$$\sigma^2_{y-target} = 0.45 + 0.0009 + 1.33 \times 10^{-4} + 8.3 \times 10^{-2}$$

which are closer to the goal. However, one must keep in mind that the inclinometer is accurate only when there is no vehicle acceleration. This is a serious restriction that must be considered. Another candidate for attitude measurement would be an inertial navigation system (INS). These are much more expensive than inclinometers, of course, but are much more accurate and can have rotational drift as low as a few milli-degrees per hour.

The application of UAVs to the problem of remote sensing as well as to other areas is rapidly expanding. The ever growing capability, coupled with the fact that the operator is not exposed to danger, makes this a popular choice for consideration.

Exercise

1   An airborne vehicle is at altitude 200 m. Assume that the vehicle is in steady flight. An object on the ground below is at coordinates $x = 70$ and $y = -60$ in vehicle coordinates. Using the instrument specifications cited in the discussion in the chapter, compute the precision one could achieve for geo-registration of this object using DGPS for positioning. Perform the attitude calculations based on GPS specifications and on inclinometer specifications and compare the overall results.

References

Newcome, Laurence R., *Unmanned Aviation: A Brief History of Unmanned Aerial Vehicles*, American Institute of Aeronautics and Astronautics, Inc., 2004.

Shim, D. H., Kim, H. J., and Sastry, S., "Hierarchical Control System Synthesis for Rotorcraft-Based Unmanned Aerial Vehicles," *AIAA Guidance, Navigation, and Control Conference and Exhibit* (AIAA, Denver, CO, August 14–17, 2000).

11

Dynamics Modeling of AUVs

11.1   Introduction

This chapter is devoted to the three-dimensional dynamics modeling of autonomous underwater vehicles (AUVs) with a focus on energy-efficient buoyancy-driven underwater gliders that are equipped with a control surface such as a rudder or an elevator. A full-order 6-DOF dynamic model is introduced, followed by the derivation of reduced-order dynamics in the longitudinal plane and three-dimensional spiraling dynamics.

11.2   Motivation

For surface vehicles motion is normally confined to the x–y plane. Thus, two-dimensional models are sufficient unless the vehicle is operated at high accelerations where even tire dynamics become important. Further, with certain assumptions the motion can be described with a simple kinematic model. While this would not be the case with a high-performance automobile or other ground vehicles, it is sufficient for many surface vehicles. Thus, kinematic models have been used in the previous chapters for describing the motion of the surface vehicles.

For underwater vehicles this is not the case since the motion is truly three dimensional. The location of the vehicle requires three dimensions, x, y, and z, and to express the attitude requires yaw, pitch, and roll. Steering is accomplished by operating control surfaces and adjusting net buoyancy whose steering effects depend also on vehicle speed. In order to deal with this more complex vehicle, this chapter investigates a full dynamic model of underwater vehicles. Certain simplifications allow for simpler models when motion is restricted to the longitudinal dimension and for other special cases.

Mobile Robots: Navigation, Control and Sensing, Surface Robots and AUVs,
Second Edition. Gerald Cook and Feitian Zhang.
© 2020 by The Institute of Electrical and Electronics Engineers, Inc.
Published 2020 by John Wiley & Sons, Inc.

Over the past decades, AUVs have attracted increasing academic and industrial attentions with a wide range of applications including, but not limited to, mapping the seafloor for oil and gas resources, searching the ocean for missing airplanes, and patrolling harbors for national security. This chapter will cover the principled dynamic modeling approach for those AUVs.

11.3 Full Dynamic Model

We model AUVs as a rigid-body system and consider an AUV with three actuation systems for locomotion including the buoyancy system, the mass distribution system, and the control surface system.

The buoyancy system changes the net buoyancy of an AUV through transferring fluid in/out of an external bladder or an internal reservoir. The mass distribution system typically uses a linear/rotary actuator to push/spin a mass (battery pack) to change the center of the mass of the AUV. The control surface system includes a rudder or an elevator for stabilization and maneuver.

Figure 11.1 shows the mass distribution of the AUV. The stationary body mass m_s (excluding the movable mass) has three components: hull mass m_h (assumed to be uniformly distributed), point mass m_w accounting for nonuniform hull mass distribution with displacement r_w with respect to the geometry center (GC), and ballast mass m_b (water in the tank) at the GC, which is a reasonable simplification since the effect on the center of gravity caused by the water in the tank is negligible compared with the effect from the movable mass. The movable mass $\overline{m}$, which is located at r_p with respect to the GC, provides a moment to the vehicle. Without loss of generality, we assume the motion of the

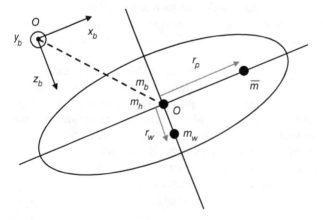

Figure 11.1 The mass distribution of an AUV (side view).

Figure 11.2 Illustration of the reference frame and hydrodynamic forces.

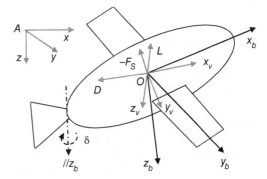

movable mass is restricted to the longitudinal axis. The vehicle hull displaces a volume of fluid of mass m. Let $m_0 = m_s + \bar{m} - m$ represent the excess mass (negative net buoyancy). The vehicle will sink if $m_0 > 0$ and ascend if $m_0 < 0$.

The relevant coordinate reference frames are defined as follows. The body-fixed reference frame, denoted as $Ox_b y_b z_b$ and shown in Figure 11.2, has its origin O at the geometry center, so the origin will be the point of application for the buoyancy force. The Ox_b axis is along the body's longitudinal axis pointing to the head; the Oz_b axis is perpendicular to Ox_b axis in the longitudinal plane of the vehicle pointing downward, and Oy_b axis is automatically formed by the right-hand orthonormal principle. In the inertial frame $Axyz$, Az axis is along gravity direction, and Ax/Ay are defined in the horizontal plane, while the origin A is a fixed point in space.

Let R represent the rotation matrix from the body-fixed reference frame to the inertial frame. R is parameterized by three Euler angles: the roll angle ϕ, the pitch angle θ, and the yaw angle ψ. Here

$$
R = \begin{pmatrix}
c\theta c\phi & s\phi s\theta c\psi - c\phi s\psi & c\phi s\theta c\psi + s\phi s\psi \\
c\theta s\psi & c\phi c\psi + s\phi s\theta s\psi & -s\phi c\psi + c\phi s\theta s\psi \\
-s\theta & s\phi c\theta & c\phi c\theta
\end{pmatrix}
\tag{11.1}
$$

where $s(\cdot)$ is short for $\sin(\cdot)$ and c for $\cos(\cdot)$. Let $v_b = [v_1\ v_2\ v_3]^T$ and $\omega_b = [\omega_1\ \omega_2\ \omega_3]^T$ represent the translational velocity and angular velocity, respectively, expressed in the body-fixed frame. The subscript b indicates that the vector is expressed in the body-fixed frame, and this notation is applied throughout this chapter.

We recognize that the rotation matrix (11.1) is different from the one defined in equation (3.5). As described in the last paragraph in Section 3.1, we follow the aerospace convention for an AUV, which differs from the one used in Chapter 3 for a ground wheeled robot. Here yaw is the angular movement in the counter clockwise direction as one looks into the z axis that points in a downward direction. Seen from above, this would be in the clockwise direction (Figure 11.2). Thus positive yaw is the opposite here than for the coordinate system used for the wheeled robot. Pitch is measured in the same way for both systems. Here it is about the y axis versus being measured about the x axis for the ground robot simply because of the different axis definitions. Nevertheless, for both coordinate systems pitch measures the upward angle of the nose. Roll here is defined as rotation about the longitudinal axis (x axis) measured in the counter clockwise direction looking into the axis. For the coordinate system used for the wheeled robot, it is the same except that there the longitudinal axis is the y axis. Thus physically, roll is the same for both coordinate systems as is also pitch. Yaw has a reversal of sign due to the opposite directions of the z axes. In addition, displacement in the z direction is opposite for the two coordinate systems. Taking these things into account, one can confirm that with the same yaw–pitch–roll rotation sequence the two rotation matrices in equations (11.1) and (3.5) are consistent in their properties and each is correct.

We assume that the AUV is equipped with a control surface—a rudder/an elevator that is rigid and pivots about the Oz_b/Oy_b axis. There are hydrodynamic forces and moments generated because of the relative movement between the control surface and the surrounding water, like the side/lift force and the yaw/pitch moment.

The dynamic model for the AUV with a control surface is as follows:

$$\dot{b}_i = R v_b \tag{11.2}$$

$$\dot{R} = R \hat{\omega}_b \tag{11.3}$$

$$\dot{v}_b = M^{-1}\left(M v_b \times \omega_b + m_0 g R^T k + F_{ext}\right) \tag{11.4}$$

$$\dot{\omega}_b = J - \left(-\dot{J}\omega_b + J\omega_b \times \omega_b + M v_b \times v_b + T_{ext} + m_w g r_w \times \left(R^T k\right) + \overline{m} g r_p \times \left(R^T k\right)\right) \tag{11.5}$$

Here $M = (m_s + \overline{m})I + M_f = \text{diag}\{m_1, m_2, m_3\}$, where I is the 3×3 identity matrix, and M_f is the added-mass matrix, which can be calculated via strip theory (Milgram, 2007). $J = \text{diag}\{J_1, J_2, J_3\}$ is the sum of the inertia matrix due to the stationary mass distribution and the added inertia matrix in water. In addition, k is the unit vector along the Az direction in the inertial frame, r_w is a constant vector, and r_p is the controllable movable mass position vector, which has one degree of freedom in the Ox_b direction, $r_p = [r_{p1} \ 0 \ 0]^T$. $b_i = [x \ y \ z]^T$ is the position vector of the vehicle in the inertial reference frame. $\hat{\omega}_b$ is the skew-symmetric matrix corresponding to ω_b. The operator $\hat{\ }$ transforms a vector a into its equivalent matrix and is used to represent the cross product, i.e., $\hat{a}b = a \times b$. F_{ext} stands for all hydrodynamic forces (lift force, drag force, and side force) acting on the AUV, expressed in the body-fixed frame. Finally, T_{ext} is the total hydrodynamic moment caused by F_{ext}.

11.4 Hydrodynamic Model

In order to model the hydrodynamics, we first introduce the velocity reference frame $Ox_v y_v z_v$. Ox_v axis is along the direction of the velocity, and Oz_v lies in the longitudinal plane perpendicular to Ox_v. Rotation matrix $\boldsymbol{R}_{bv}$ represents the rotation operation from the velocity reference frame to the body-fixed frame:

$$R_{bv} = \begin{pmatrix} c\alpha c\beta & -c\alpha s\beta & -s\alpha \\ s\beta & c\beta & 0 \\ s\alpha c\beta & -s\alpha s\beta & c\alpha \end{pmatrix} \tag{11.6}$$

where the angle of attack $\alpha = \arctan(v_3/v_1)$ and the sideslip angle $\beta = \arcsin(v_2/\|\boldsymbol{v}_b\|)$.

The hydrodynamic forces include the lift force L, the drag force D, and the side force F_S; the hydrodynamic moments include the roll moment M_1, the pitch moment M_2, and the yaw moment M_3. All of those forces and moments are defined in the velocity frame (Panton, 2005). And if we further assume that the control surface deflects slowly and smoothly, usually true for the AUV stabilization and maneuverability, the propelling force from the control surface will be negligible compared to the buoyancy-induced propelling force. Then we will have the following relationship:

$$F_{ext} = R_{bv}\left[-D \quad F_S \quad -L\right]^T \tag{11.7}$$

$$T_{ext} = R_{bv}\left[M_1 \quad M_2 \quad M_3\right]^T \tag{11.8}$$

The hydrodynamic forces and moments are dependent on the angle of attack α, the sideslip angle β, the velocity magnitude V, and the control surface angle δ (Anderson, 1998). Here we consider the rudder for the control surface as an example:

$$D = 1/2\rho V^2 S\left(C_{D0} + C_D^\alpha \alpha^2 + C_D^\delta \delta^2\right) \tag{11.9}$$

$$F_S = 1/2\rho V^2 S\left(C_{F_S}^\beta \beta + C_{F_S}^\delta \delta\right) \tag{11.10}$$

$$L = 1/2\rho V^2 S\left(C_{L0} + C_L^\alpha \alpha\right) \tag{11.11}$$

$$M_1 = 1/2\rho V^2 S\left(C_{M_R}^\beta \beta + K_{q1}\omega_1\right) \tag{11.12}$$

$$M_2 = 1/2\rho V^2 S\left(C_{M_0} + C_{M_P}^\alpha \alpha + K_{q2}\omega_2\right) \tag{11.13}$$

$$M_3 = 1/2\rho V^2 S\left(C_{M_Y}^\beta \beta + K_{q3}\omega_3 + C_{M_Y}^\delta \delta\right) \tag{11.14}$$

where ρ is the density of water and S is the characteristic area of the AUV. The control surface (rudder) angle δ is defined as the angle between the longitudinal axis Ox_b and the center line of the rudder projected into the $Ox_b y_b$ plane, with Oz_b axis as the positive direction. K_{q1}, K_{q2}, K_{q3} are rotation damping coefficients. All other constants with "C" in their notations are hydrodynamic coefficients, whose values can be evaluated through theoretical calculation, Computational Fluid Dynamics (CFD) simulation, or towing tank experiments.

We recognize that the hydrodynamic model of underwater vehicles shares great similarity with the aerodynamic model of aerial vehicles. For example, in both air and water, the drag force D is quadratic in the angle of attack α and the lift force L is linear in α. While the scales of hydrodynamic/aerodynamic coefficients may differ significantly, we expect our study will provide some fundamental modeling and control knowledge for both underwater and aerial vehicles.

11.5 Reduced-Order Longitudinal Dynamics

Steady-state gliding in the longitudinal plane is one of the most important operations for underwater gliders. When the motion of the vehicle is restricted to the longitudinal plane, we have

$$R = \begin{pmatrix} \cos\theta & 0 & \sin\theta \\ 0 & 1 & 0 \\ -\sin\theta & 0 & \cos\theta \end{pmatrix}, \quad b_i = \begin{pmatrix} x \\ 0 \\ z \end{pmatrix}, \quad v_b = \begin{pmatrix} v_1 \\ 0 \\ v_3 \end{pmatrix}$$

$$\omega_b = \begin{pmatrix} 0 \\ \omega_2 \\ 0 \end{pmatrix}, \quad r_p = \begin{pmatrix} r_{p1} \\ 0 \\ 0 \end{pmatrix}, \quad r_w = \begin{pmatrix} 0 \\ 0 \\ r_{w3} \end{pmatrix}, \quad \delta = 0$$

Here we assume that the point mass m_w is just below the center of geometry O by r_{w3} as such bottom-heavy design is desirable for stability concern and also achievable with manufacture. Plugging the hydrodynamic forces and moments into the vehicle dynamics equations, we get the following model:

$$\dot{v}_1 = (m_1 + \bar{m})^{-1}(-(m_3 + \bar{m})v_3\omega_2 - m_0 g \sin\theta + L \sin\alpha - D\cos\alpha) \tag{11.15}$$

$$\dot{v}_3 = (m_3 + \bar{m})^{-1}((m_1 + \bar{m})v_1\omega_2 + m_0 g \cos\theta - L\cos\alpha - \sin\alpha) \tag{11.16}$$

$$\dot{\omega}_2 = J_2^{-1}\left(M_2 + (m_3 - m_1)v_1 v_3 - m_w g r_{w3} \sin\theta - \bar{m}g r_{p1} \cos\theta\right) \tag{11.17}$$

$$\dot{x} = v_1 \cos \theta + v_3 \sin \theta \qquad (11.18)$$

$$\dot{z} = -v_1 \sin \theta + v_3 \cos \theta \qquad (11.19)$$

$$\dot{\theta} = \omega_2 \qquad (11.20)$$

We choose a miniature underwater glider to demonstrate the gliding motion governed by the derived reduced-order longitudinal dynamics. The hydrodynamic coefficients of the selected vehicle are

$$C_{D_0} = 0.45275, \quad C_D^\alpha = 17.5948$$
$$C_{L_0} = 0.074606, \quad C_L^\alpha = 19.5777$$
$$C_{M_0} = 0.0075719, \quad C_{M_P}^\alpha = 0.5665$$

Figures 11.3 and 11.4 show the control inputs used in the simulation. Movable mass displacement and net buoyancy take the form of square waves and the rudder angle is kept at $0°$. The achieved zigzag gliding path is illustrated in Figure 11.5. The trajectories of the pitch angle and the speed of the vehicle in the longitudinal-plane gliding are shown in Figures 11.6 and 11.7. We observe from simulation results that for fixed control inputs, the system states and the vehicle path converge to the steady state after a short period of transient dynamics. This zigzag gliding pattern is the most common operating motion of underwater gliders, which takes advantages of gravity and buoyancy for energy-efficient locomotion.

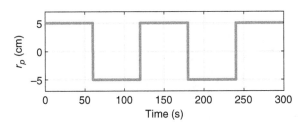

Figure 11.3 The control input plot of the movable mass displacement in the longitudinal gliding simulation.

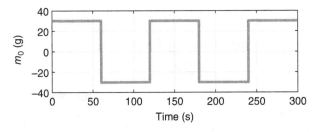

Figure 11.4 The control input plot of the net buoyancy in the longitudinal gliding simulation.

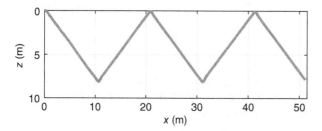

Figure 11.5 The zigzag gliding path in the longitudinal gliding simulation.

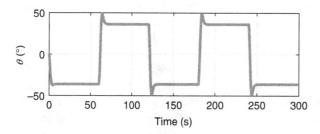

Figure 11.6 Plot of the pitch angle in the longitudinal gliding simulation.

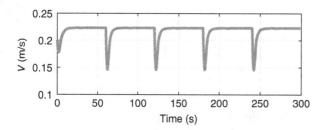

Figure 11.7 Plot of the gliding speed in the longitudinal gliding simulation.

11.6 Computation of Steady Gliding Path in the Longitudinal Plane

During steady glide, the angular velocity is zero, while the translational velocity stays unchanged. The control r_{p1} and m_0 are constant, which means that the position of the movable mass is fixed with respect to the origin O and the pumping rate is zero. So the steady motion can be described by the following algebraic equations:

$$0 = -m_0 g \sin\theta + L \sin\alpha - D \cos\alpha \tag{11.21}$$

$$0 = m_0 g \cos\theta - L \cos\alpha - D \sin\alpha \tag{11.22}$$

$$0 = M_2 + (m_3 - m_1)v_1 v_3 - m_w g r_{w3} \sin\theta - \bar{m} g r_{p1} \cos\theta \tag{11.23}$$

In addition, we recall the definition of the angle of attack α as a function of the velocity components v_1 and v_3, i.e.,

$$\alpha = \arctan(v_3/v_1) \tag{11.24}$$

The solution to the above equations gives us the steady gliding path.

We choose the same miniature underwater glider in Section 11.5 to demonstrate the computation of steady gliding path. As a comparative trial, another set of wings is used with the same wingspan but doubled aspect ratio (i.e., chord length is half), while the vehicle body is left unchanged. The hydrodynamic coefficients for the miniature underwater glider with smaller wings are

$$C_{D_0} = 0.44724, \quad C_D^\alpha = 10.298$$
$$C_{L_0} = 0.054273, \quad C_L^\alpha = 11.5545$$
$$C_{M_0} = 0.0062683, \quad C_{M_P}^\alpha = 0.2903$$

With the hydrodynamic parameters obtained, let us take a look at the solution of the steady gliding equations (11.21)–(11.23). These equations are highly nonlinear due to the terms involving trigonometric functions and inverse trigonometric functions in the state. When the angle of attack is small enough, we can use the approximation $\sin\alpha \approx \alpha$ and $\cos\alpha \approx 1$, and derive an approximate analytical solution for the desired control r_p and m_0 in order to achieve some given steady states (Mahmoudian et al., 2010). However, here we are interested in the calculation of the steady gliding states themselves under a fixed control. Unfortunately, there are no feasible analytical solutions for this problem. With Matlab command *solve()*, we numerically solve equations (11.21)–(11.23) to get the velocity $\mathbf{v}$, pitch angle θ, and glide angle θ_g for a given movable mass displacement r_p and net buoyancy m_0, under different conditions for r_w, the location of nonuniform stationary mass. There is only one feasible solution for each pair of (r_p, m_0). Other solutions are rejected based on their physical interpretations.

Tables 11.1 and 11.2 show scan results where the steady gliding path is presented with different sets of center of mass distribution, location of movable mass, and net buoyancy, for two different wing designs—a larger pair (Table 11.1) and a smaller pair (Table 11.2) of wings. The gliding angle $\theta_g = \theta - \alpha$ is the angle between Ox_v and Ax with gliding up as positive. z_{CG} stands for

Table 11.1 Computed steady gliding path under different values of the center of gravity z_{CG}, the movable mass displacement r_{p1}, and the excess mass m_0, for the underwater glider model with larger wings.

z_{CG} (cm)	r_{p1} (cm)	m_0 (g)	(V, α, θ_g) (m/s, °, °)
0.1	0.3	10	(0.1129 3.0470 −29.5522)
0.1	0.5	10	(0.1366 1.6543 −43.0404)
0.1	0.7	10	(0.1485 1.0936 −52.7389)
0.1	0.3	30	(0.1766 3.9483 −25.0827)
0.1	0.5	30	(0.2245 2.0106 −38.4395)
0.1	0.7	30	(0.2495 1.3300 −48.7594)
0.1	0.3	50	(0.2245 4.0967 −24.5276)
0.1	0.5	50	(0.2846 2.1371 −37.0375)
0.1	0.7	50	(0.3174 1.3980 −47.0516)
0.2	0.3	10	(0.0856 5.8988 −20.2069)
0.2	0.5	10	(0.1084 3.3827 −27.6331)
0.2	0.7	10	(0.1240 2.3211 −35.1835)

Table 11.2 Computed steady gliding path under different values of the center of gravity z_{CG}, the movable mass displacement r_{p1}, and the excess mass m_0, for the underwater glider model with smaller wings.

z_{CG} (cm)	r_{p1} (cm)	m_0 (g)	(V, α, θ_g) (m/s, °, °)
0.1	0.3	10	(0.1221 4.0658 −37.0187)
0.1	0.5	10	(0.1396 2.4940 −48.3662)
0.1	0.7	10	(0.1486 1.7575 −56.4820)
0.1	0.3	30	(0.2260 3.2732 −41.9002)
0.1	0.5	30	(0.2477 2.2108 −51.2303)
0.1	0.7	30	(0.2598 1.6385 −58.0061)
0.1	0.3	50	(0.3105 2.5525 −47.8110)
0.1	0.5	50	(0.3290 1.8747 −55.0414)
0.1	0.7	50	(0.3401 1.4598 −60.4123)
0.2	0.3	10	(0.0949 7.7979 −26.1314)
0.2	0.5	10	(0.1136 5.0110 −32.7957)
0.2	0.7	10	(0.1265 3.6361 −39.4833)

the center of gravity expressed in the z-axis coordinate of the body-fixed frame and there is a bijective function from r_{w3} to z_{CG}:

$$z_{CG} = \frac{m_w}{m} r_{w3} \tag{11.25}$$

Here we ignore the influence of the excess mass m_0 on z_{CG}, which is really small compared to that of m_w. From the results, we can see that different wing designs lead to different static gliding profiles. For example, the larger wings result in shallower gliding paths (longer horizontal travel) but slower total speed compared to the smaller wings, given the same set of control inputs. Since typically the wings can be easily replaced in the design, we can potentially tailor the wing designs, while leaving the glider body and its inside intact, to accommodate the requirements of different applications. On the other hand, the results in the table show that, for a fixed wing design, the speed is influenced by both the excess mass m_0 and the pair (r_p, z_{CG}) while the pitch angle is affected mainly by the pair (r_p, z_{CG}). Therefore, the center of mass plays an important role in determining the steady gliding attitude. In particular, if we compare the cases where the values of z_{CG} are different but the other parameters are the same, we find that smaller z_{CG} results in higher speed and larger glide angle. This observation can be used in the design—by making z_{CG} small, one can achieve the desired glide angle with very small displacement of the movable mass.

11.7 Scaling Analysis

We study the larger-wing glider model at different scales and introduce a new cost performance index, which reflects the horizontal travel distance per unit energy consumption. For one dive (descent and ascent), the horizontal travel distance D_d is approximated as

$$D_d = V_h t_d = 2\frac{V_h h}{V_v} \tag{11.26}$$

where V_h and V_v are the steady-state horizontal speed and vertical speed, respectively, t_d is the travel time for one dive, and h is the vertical travel depth. The energy consumption in one dive comes from two sources, the pump actuation and the movable mass actuation, while the energy consumed for the latter is negligible compared to that for pumping since the pump needs to overcome large pressure when the glider switches to ascent from descent. So the energy consumption per dive E_d can be approximated as

$$E_d = \rho g h_0 S_p l_p + \rho g (h_0 + h) S_p l_p \tag{11.27}$$

Here, ρ is the water density, h_0 is the equivalent water depth of the atmosphere pressure, S_p is the cross-section area of the pump tank inlet (and outlet), and l_p

represents the length of the water column if the water pumped in each cycle is placed in a cylindrical container with cross-section area S_p. Noting the net buoyancy $m_0 = \frac{1}{2}\rho S_p l_p$, we further simplify the energy consumption per dive to $E_d = 2m_0 g(2h_0 + h)$. Then we have the horizontal travel distance per unit energy consumption

$$\frac{D_d}{E_d} = \frac{V_h}{V_v m_0} \frac{1}{1 + 2(h_0/h)} \tag{11.28}$$

For a specific task, the depth is fixed and we have

$$\frac{D_d}{E_d} \propto \frac{V_h}{V_v m_0} \tag{11.29}$$

Therefore, we define the cost performance index τ as follows

$$\tau = \frac{V_h}{V_v m_0} \tag{11.30}$$

Here a larger τ indicates better energy efficiency, and an optimal design/control strategy is to maximize the performance index over the vehicle design/state space.

We now conduct scaling analysis to examine how the cost performance metric evolves with the dimensions (and consequently the weight) of the glider. CFD simulation shows that the drag coefficient C_D and lift coefficient C_L stay almost the same at different scales we considered (from $0.25:1$ to $8:1$), while the pitch moment coefficient C_M scales linearly with the characteristic dimension l of the glider. All related masses of the glider will scale as l^3, including the movable mass $\bar{m}$ and the negative net buoyancy m_0. Taking the total length of the glider as l, the scale $1:1$ would imply $l = 50$ cm. By plugging those new parameter values into equations (11.21)–(11.23), we can solve the glide path for the scaled model.

Table 11.3 shows the glide paths for glider models at different scales. Figure 11.8 shows the relationship between the cost performance index $\frac{V_h}{V_v v_m}$ and the scale. The results show us that with a larger body the glider has a smaller glide ratio and a smaller cost performance index value, thus consuming more energy for a given horizontal travel distance. This is consistent with the fact that a larger glider needs to pump more water for a proper net buoyancy to provide the propelling force, which is also the main energy consumption source. However, a larger-scale glider is able to achieve faster horizontal speed as shown in Figure 11.9. There is a trade-off between the achieved horizontal speed and the horizontal distance coverage per unit energy cost, when selecting the optimal scale for the glider. Other factors, like the dimension and the mass of the sensors and actuation devices, should be also taken into account in the design process.

Table 11.3 Computed steady gliding path for the scaled models of the larger wing underwater glider. In computation, $r_p = 5$ mm is used for the original scale model (1 : 1) while the value is scaled linearly with dimension for other models.

Scale	Mass (kg)	m_0 (kg)	V_h (m/s)	V_v (m/s)	$\dfrac{V_h}{V_v m_0}$ (kg^{-1})	Glide ratio
0.25 : 1	1	0.0075	0.063	0.018	488.35	3.5
0.5 : 1	2	0.015	0.11	0.039	203.25	2.82
1 : 1	4	0.03	0.19	0.094	74.55	2.02
2 : 1	8	0.06	0.28	0.207	28.30	1.35
4 : 1	16	0.12	0.39	0.377	12.01	1.03
8 : 1	32	0.24	0.54	0.574	5.72	0.94

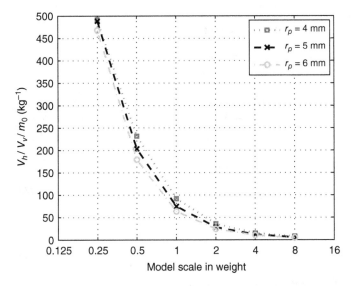

Figure 11.8 The glider cost performance index with respect to model scales.

11.8 Spiraling Dynamics

If control inputs are fixed with nonzero rudder angle, we can treat the influence of the rudder on the hydrodynamic forces and moments as the effects of increased hydrodynamic angles (α, β), and we know that the underwater glider will perform three-dimensional steady spiraling motion (Zhang et al., 2014), where the yaw angle ψ changes at constant rate while the roll angle ϕ and pitch angle θ are constants. Then $R^T k$ is constant since

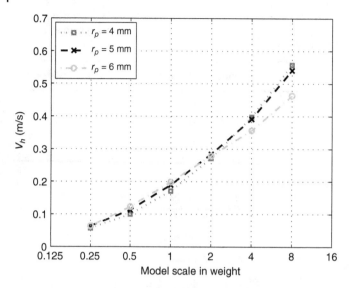

Figure 11.9 The horizontal velocity with respect to model scales.

$$R^T k = R^T \begin{pmatrix} 0 \\ 0 \\ 1 \end{pmatrix} = \begin{pmatrix} -\sin\theta \\ \sin\phi\cos\theta \\ \cos\phi\cos\theta \end{pmatrix} \tag{11.31}$$

Taking time derivative of $R^T k$, we have

$$\boldsymbol{\omega}_b \times \left(R^T k \right) = 0 \tag{11.32}$$

so the angular velocity has only one degree of freedom with ω_{3i} in Oz axis in the inertial frame. Then

$$\boldsymbol{\omega}_b = \omega_{3i} \left(R^T k \right) \tag{11.33}$$

The translational velocity in the body-fixed frame

$$v_b = R_{bv}[V \quad 0 \quad 0]^T \tag{11.34}$$

There are two important parameters in the spiraling motion: the turning radius R and the vertical speed $V_{vertical}$. By projecting the total velocity into the horizontal plane and vertical direction, we have

$$V_{vertical} = R_{bv}(V \quad 0 \quad 0)^T \left(R^T k \right) \tag{11.35}$$

$$R = \sqrt{V^2 - V_v^2} \Big/ \omega_{3i} \tag{11.36}$$

The steady-state spiraling equations are obtained by setting time derivatives to zero for the vehicle dynamics:

$$0 = M\mathbf{v}_b \times \boldsymbol{\omega}_b + m_0 g \mathbf{R}^T \mathbf{k} + \mathbf{F}_{ext} \tag{11.37}$$

$$0 = \mathbf{J}\boldsymbol{\omega}_b \times \boldsymbol{\omega}_b + M\mathbf{v}_b \times \mathbf{v}_b + \mathbf{T}_{ext} + m_w g \mathbf{r}_w \times \left(\mathbf{R}^T \mathbf{k}\right) + \bar{m} g \mathbf{r}_p \times \left(\mathbf{R}^T \mathbf{k}\right) \tag{11.38}$$

From equations (11.1), (11.6), (11.33), and (11.34) and the above steady-state spiraling equations, we know there are six independent states for describing the steady spiral motion: $[\theta \ \phi \ \omega_{3i} \ V \ \alpha \ \beta]$ with $[m_0 \ r_{p1} \ \delta]$ as the three control inputs. Expanding equations (11.37) and (11.38), and then transforming the original states to the above six independent states, we can obtain the nonlinear steady-state spiraling equations as in (11.39)–(11.44).

$$\begin{aligned}
0 = {}& m_2 s\beta V c\phi c\theta \omega_{3i} - m_3 s\alpha c\beta V \ s\phi c\theta \omega_{3i} - m_0 g s\theta \\
& - 1/2\rho V^2 S\left(C_{SF}^\beta \beta + C_{SF}^\delta \delta\right) c\alpha s\beta + 1/2\rho V^2 S\left(C_{L0} + C_L^\alpha \alpha\right) s\alpha \\
& - 1/2\rho V^2 S\left(C_{D0} + C_D^\alpha \alpha^2 + C_D^\delta \delta^2\right) c\alpha c\beta
\end{aligned} \tag{11.39}$$

$$\begin{aligned}
0 = {}& -m_3 s\alpha c\beta V s\theta \omega_{3i} - m_1 c\alpha c\beta V \ c\phi c\theta \omega_{3i} \\
& - 1/2\rho V^2 S\left(C_{D0} + C_D^\alpha \alpha^2 + C_D^\delta \delta^2\right) s\beta + m_0 g s\phi c\theta \\
& + 1/2\rho V^2 S\left(C_{SF}^\beta \beta + C_{SF}^\delta \delta\right) c\beta
\end{aligned} \tag{11.40}$$

$$\begin{aligned}
0 = {}& m_1 c\alpha c\beta V s\phi c\theta \omega_{3i} + m_2 s\beta V s\theta \omega_{3i} + m_0 g c\phi c\theta \\
& - 1/2\rho V^2 S\left(C_{SF}^\beta \beta + C_{SF}^\delta \delta\right) s\alpha s\beta - 1/2\rho V^2 S\left(C_{L0} + C_L^\alpha \alpha\right) c\alpha \\
& - 1/2\rho V^2 S\left(C_{D0} + C_D^\alpha \alpha^2 + C_D^\delta \delta^2\right) s\alpha c\beta
\end{aligned} \tag{11.41}$$

$$\begin{aligned}
0 = {}& (J_2 - J_3) s\phi c\theta c\phi c\theta \omega_{3i}^2 + 1/2\rho V^2 S\left(C_{M_R}^\beta \beta - K_{q1} s\theta \omega_{3i}\right) c\alpha c\beta \\
& - 1/2\rho V^2 S\left(C_{M0} + C_{M_P}^\alpha \alpha + K_{q2} s\phi c\theta \omega_{3i}\right) c\alpha s\beta \\
& - 1/2\rho V^2 S\left(C_{M_Y}^\beta \beta + K_{q3} c\phi c\theta \omega_{3i} + C_{M_Y}^\delta \delta\right) s\alpha \\
& - m_w g r_w s\phi c\theta + (m_2 - m_3) s\beta s\alpha c\beta V^2
\end{aligned} \tag{11.42}$$

$$\begin{aligned}
0 = {}& (J_1 - J_3) s\theta c\phi c\theta \omega_{3i}^2 + (m_3 - m_1) c\alpha c\beta s\alpha c\beta V^2 - m_w g r_w s\theta \\
& + 1/2\rho V^2 S\left(C_{M_R}^\beta - K_{q1} s\theta \omega_{3i}\right) s\beta - \bar{m} g r_{p1} c\phi c\theta \\
& + 1/2\rho V^2 S\left(C_{M_0} + C_{M_P}^\alpha \alpha + K_{q2} s\phi c\theta \omega_{3i}\right) c\beta
\end{aligned} \tag{11.43}$$

$$
\begin{aligned}
0 = {} & (J_2 - J_1)s\theta s\phi c\theta\omega_{3i}^2 + (m_1 - m_2)cac\beta s\beta V^2 \\
& - 1/2\rho V^2 S\left(C_{M_0} + C_{M_p}^\alpha \alpha + K_{q2}s\phi c\theta\omega_{3i}\right)sas\beta + \bar{m}gr_{p1}s\phi c\theta \\
& + 1/2\rho V^2 S\left(C_{M_Y}^\beta \beta + K_{q3}c\phi c\theta\omega_{3i} + C_{M_Y}^\delta \delta\right)ca \\
& + 1/2\rho V^2 S\left(C_{M_R}^\beta \beta - K_{q1}s\theta\omega_{3i}\right)sac\beta
\end{aligned}
$$

$$(11.44)$$

Here, we assume the mass matrix and inertia matrix have the following form:

$$
M = \begin{pmatrix} m_1 & 0 & 0 \\ 0 & m_2 & 0 \\ 0 & 0 & m_3 \end{pmatrix} \quad J = \begin{pmatrix} J_1 & 0 & 0 \\ 0 & J_2 & 0 \\ 0 & 0 & J_3 \end{pmatrix}
$$

We take the same miniature underwater glider from Section 11.5 as an example to demonstrate the spiral dynamics. With all three control inputs fixed, specifically, with the movable mass displacement at 5 cm, the net buoyancy at 30 g, and the rudder angle at $45°$, we simulate the vehicle dynamics. Figure 11.10 shows the three-dimensional helical spiraling path of the underwater glider. Figures 11.11–11.13 show the trajectories of the three Euler angles. We confirm

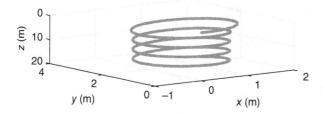

Figure 11.10 The helical vehicle path in the spiral dynamics simulation.

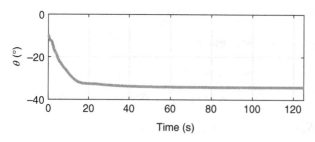

Figure 11.11 Plot of the pitch angle in the spiral dynamics simulation.

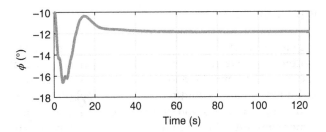

Figure 11.12 Plot of the roll angle in the spiral dynamics simulation.

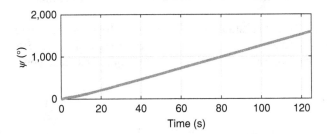

Figure 11.13 Plot of the yaw angle in the spiral dynamics simulation.

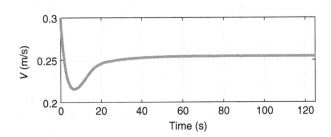

Figure 11.14 Plot of the vehicle speed in the spiral dynamics simulation.

from simulation results that the pitch and roll converge to constant values in the steady spiral while the yaw angle changes at a constant rate. The spiraling speed of the vehicle is also a constant as shown in Figure 11.14. This spiral motion of underwater gliders is typically used in sampling water columns due to its energy-efficient descending/ascending dynamics and the unique three-dimensional helical path with controllable spiral radius.

11.9 Computation of Spiral Path

The steady-state spiraling equations are highly nonlinear due to the terms involving trigonometric functions and inverse trigonometric functions. Given the angle of attack α, the sideslip angle β, and the velocity magnitude V, a recursive algorithm based on fixed-point iteration could potentially be applied to solve the equations for the other system states and control inputs (Zhang et al., 2013). However, we are more interested in the converse problem of how to calculate steady-state solutions given fixed control inputs, which are more useful for path planning and control purposes. Unfortunately, this problem does not admit analytical solutions and the convergence condition for the corresponding fixed-point problem is not satisfied. In the following, we apply Newton's method to solve the problem.

Let $x = [\theta \; \phi \; \omega_{3i} \; V \; \alpha \; \beta]^T$ be the six states that we want to solve for steady-state spiral gliding equations. And let $u = [m_0 \; r_{p1} \; \delta]^T$ be the three control inputs. For convenience of presentation, we write the governing equations in a compact form

$$0 = f(x,u) = [f_i(x,u)]_{i=1,\ldots,6} \tag{11.45}$$

For example, f_1 is the right-hand side of equation (11.39).

The iterative algorithm for Newton's method reads (Kelley, 2003)

$$\hat{x}_{i+1} = \hat{x}_i - J^{-1}(\hat{x}_i, u) f(\hat{x}_i, u) \tag{11.46}$$

Here $\hat{x}_i$ is the ith-step iteration for the steady states, and $J(x, u)$ is the Jacobian matrix of $f(x, u)$

$$J(x,u) = \frac{\partial f}{\partial x} = \left[\frac{\partial f_i}{\partial x_j}\right]_{6 \times 6} \tag{11.47}$$

The first row elements of the Jacobian matrix are given in equations (11.48)–(11.53) while the others are omitted for succinct presentation, which can be calculated similarly.

$$\partial f_1/\partial x_1 = -m_2 s\beta V c\phi s\theta \omega_{3i} + m_3 s\alpha c\beta V s\phi s\theta \omega_{3i} - m_0 g c\theta \tag{11.48}$$

$$\partial f_1/\partial x_2 = -m_2 s\beta V s\phi c\theta \omega_{3i} - m_3 s\alpha c\beta V c\phi c\theta \omega_{3i} \tag{11.49}$$

$$\partial f_1/\partial x_3 = m_2 s\beta V c\phi c\theta - m_3 s\alpha c\beta V s\phi c\theta \tag{11.50}$$

$$\begin{aligned}
\partial f_1/\partial x_4 = \; & m_2 s\beta c\phi c\theta \omega_{3i} - m_0 g s\theta + \rho V S\left(C_{L0} + C_L^\alpha \alpha\right) s\alpha \\
& - \rho V S\left(C_{SF}^\beta \beta + C_{SF}^\delta \delta\right) c\alpha s\beta - m_3 s\alpha c\beta s\phi c\theta \omega_{3i} \\
& - \rho V S\left(C_{D0} + C_D^\alpha \alpha^2 + C_D^\delta \alpha^2\right) c\alpha s\beta
\end{aligned} \tag{11.51}$$

$$\partial f_1/\partial x_5 = -m_3 cac\beta V s\phi c\theta \omega_{3i} - \rho V^2 S C_D^\alpha acac\beta$$
$$+ 1/2\rho V^2 S\left(C_{SF}^\beta \beta + C_{SF}^\delta \delta\right) sas\beta + 1/2\rho V^2 S C_L^\alpha sa$$
$$+ 1/2\rho V^2 S\left(C_{D0} + C_D^\alpha \alpha\right) c\alpha \qquad (11.52)$$
$$+ 1/2\rho V^2 S\left(C_{D0} + C_D^\alpha \alpha + C_D^\delta \delta^2\right) sac\beta$$

$$\partial f_1/\partial x_6 = m_2 c\beta V c\phi c\theta \omega_{3i} - 1/2\rho V^2 S\left(C_{SF}^\beta \beta + C_{SF}^\delta \delta\right) cac\beta$$
$$+ 1/2\rho V^2 S\left(C_{D0} + C_D^\alpha \alpha^2 + C_D^\delta \delta^2\right) cas\beta \qquad (11.53)$$
$$+ m_3 sas\beta V s\phi c\theta \omega_{3i} - 1/2\rho V^2 S C_{SF}^\beta cas\beta$$

Based on the parameters of the miniature underwater glider as listed in Table 11.4, Newton's iterative formula is used to solve the steady-state spiraling equations. Characteristic parameters for steady spiraling motion, including the turning radius and ascending/descending speed, are obtained with different inputs as shown in Table 11.5. To apply Newton's method, the initial values of states for iteration are chosen to be $\theta = -10°$, $\phi = -10°$, $\omega_{3i} = 0.1$ rad/s, $V = 0.3$ m/s, $\alpha = 0°$, and $\beta = 0°$. From the calculated results, we can see that a small turning radius requires a large rudder angle, a large displacement of movable mass, and a small net buoyancy, while a low descending or ascending speed

Table 11.4 Parameters of the miniature underwater glider used in the steady-state spiraling equations.

Parameter	Value	Parameter	Value
m_1	3.88 kg	m_2	9.9 kg
m_3	5.32 kg	$\bar{m}$	0.8 kg
C_{D0}	0.45	C_D^α	17.59 rad^{-2}
$C_{F_S}^\beta$	-2 rad^{-1}	$C_{F_S}^\delta$	1.5 rad^{-1}
C_{L0}	0.075	C_L^α	19.58 rad^{-1}
J_1	0.8 kg·m^2	J_2	0.05 kg·m^2
J_3	0.08 kg·m^2	C_{M0}	0.0076 m
$C_{M_R}^\beta$	-0.3 m/rad	$C_{M_P}^\alpha$	0.57 m/rad
$C_{M_Y}^\beta$	5 m/rad	$C_{M_Y}^\delta$	-0.2 m/rad
K_{q1}	-0.1 m·s/rad	K_{q2}	-0.5 m·s/rad
K_{q3}	-0.1 m·s/rad	S	0.012 m^2

Table 11.5 Computed spiraling steady states through Newton's method.

m_0 (g)	r_{p1} (cm)	$\delta(°)$	$(\theta, \phi, \omega_{3i}, V, \alpha, \beta)$ (°, °, rad/s, m/s, °, °)	$(V_{vertical}, R)$ (m/s, m)
25	0.3	45	(−44.5, −31.0, 0.425, 0.264, −0.914, 4.10)	(0.182, 0.450)
25	0.4	45	(−46.8, −36.6, 0.448, 0.267, −1.32, 4.52)	(0.190, 0.417)
25	0.5	45	(−48.3, −40.6, 0.464, 0.268, −1.61, 4.87)	(0.195, 0.396)
25	0.6	45	(−49.3, −43.8, 0.476, 0.267, −1.84, 5.18)	(0.197, 0.380)
25	0.7	45	(−50.2, −46.5, 0.486, 0.267, −2.04, 5.48)	(0.211, 0.338)
10	0.5	45	(−70.8, −49.3, 0.589, 0.184, −3.64, 7.36)	(0.169, 0.121)
15	0.5	45	(−63.5, −52.7, 0.571, 0.218, −3.30, 6.98)	(0.189, 0.190)
20	0.5	45	(−55.5, −47.8, 0.517, 0.247, −2.46, 5.85)	(0.197, 0.287)
30	0.5	45	(−42.1, −34.3, 0.423, 0.281, −0.901, 4.24)	(0.185, 0.500)
35	0.5	45	(−36.9, −29.3, 0.392, 0.289, −0.306, 3.85)	(0.172, 0.591)
40	0.5	45	(−32.3, −25.3, 0.368, 0.293, 0.224, 3.60)	(0.157, 0.670)
25	0.5	30	(−37.6, −11.9, 0.235, 0.242, 0.854, 2.19)	(0.151, 0.806)
25	0.5	35	(−43.4, −20.7, 0.311, 0.258, 0.0698, 2.87)	(0.178, 0.602)
25	0.5	40	(−46.8, −31.2, 0.389, 0.266, −0.761, 3.77)	(0.192, 0.474)
25	0.5	50	(−49.2, −48.8, 0.537, 0.264, −2.54, 6.19)	(0.192, 0.337)
25	0.5	55	(−51.1, −56.4, 0.615, 0.257, −3.62, 7.86)	(0.190, 0.283)
25	0.5	60	(−55.0, −63.8, 0.705, 0.247, −4.95, 10.0)	(0.189, 0.225)

demands a small rudder angle, a small displacement of movable mass, and a medium net buoyancy.

Exercises

1   Consider an underwater glider with the following system parameters:

$$C_{D_0} = 0.5 \quad C_D^\alpha = 15\,\text{rad}^{-2} \quad C_{L_0} = 0.1 \quad C_L^\alpha = 20\,\text{rad}^{-1}$$
$$C_{M_0} = 0\,\text{m} \quad C_{M_p}^\alpha = 1\,\text{m/rad} \quad S = 0.1\,\text{m}^2 \quad m_w = 10\,\text{kg}$$
$$r_w = 5\,\text{cm} \quad m_1 = 10\,\text{kg} \quad m_3 = 15\,\text{kg} \quad \bar{m} = 2\,\text{kg}$$

The two control inputs include the net buoyancy m_0 and the displacement of the movable mass r_p. Compute the steady gliding paths in the longitudinal plane for all four possible control combinations of $m_0 = 0.2$ and 0.5 kg and $r_p = 10$ and 20 cm. Comment on the influences of the net buoyancy and displacement of movable mass on the steady-state gliding speed and gliding angle.

2 Consider an underwater glider equipped with a rudder control surface. The same system parameters in Exercise 1 apply as well as the following

$$C^{\beta}_{F_S} = -2\,\text{rad}^{-1} \qquad C^{\beta}_{M_R} = -0.3\,\text{m/rad} \qquad C^{\beta}_{M_Y} = 0.5\,\text{m/rad}$$

$$K_{q1} = -0.1\,\text{m·s/rad} \qquad K_{q2} = -0.5\,\text{m·s/rad} \qquad K_{q3} = -0.1\,\text{m·s/rad}$$

$$C^{\delta}_{F_S} = 1.5\,\text{rad}^{-1} \qquad C^{\delta}_{M_Y} = -0.2\,\text{m/rad} \qquad m_2 = 10\,\text{kg}$$

$$J_1 = 5\,\text{kg·m}^2 \qquad J_2 = 0.5\,\text{kg·m}^2 \qquad J_3 = 0.5\,\text{kg·m}^2$$

Two control inputs are kept constant with the net buoyancy $m_0 = 0.5\,\text{kg}$ and the displacement of the movable mass $r_p = 10\,\text{cm}$. Simulate the transient and steady-state spiral path under rudder angles $\delta = 15°$, $30°$, and $45°$, respectively. Choose the initial condition as that of steady gliding when the rudder angle is zero. Comment on the influences of the rudder angle on the steady-state spiral path.

References

Anderson, J. D., *Aircraft Performance and Design*, McGraw-Hill, New York, 1998.

Kelley, C. T., *Solving Nonlinear Equations with Newton's Method*, Society for Industrial Mathematics, 2003.

Mahmoudian, N., Geisbert, J., and Woolsey, C., "Approximate Analytical Turning Conditions for Underwater Gliders: Implications for Motion Control and Path Planning," *IEEE Journal of Oceanic Engineering*, Vol. **35**, No. 1 (2010), pp. 131–143.

Milgram, J. H., "Strip Theory for Underwater Vehicles in Water of Finite Depth," *Journal of Engineering Mathematics*, Vol. **58**, No. 1 (2007), pp. 31–50.

Panton, R. L., *Incompressible Flow*, Wiley, New York, 2005.

Zhang, F., Zhang, F., and Tan, X., "Tail-Enabled Spiraling Maneuver for Gliding Robotic Fish," *Journal of Dynamic Systems, Measurement, and Control*, Vol. **136**, No. 4 (2014), 041028.

Zhang, S., Yu, J., Zhang, A., and Zhang, F., "Spiraling Motion of Underwater Gliders: Modeling, Analysis, and Experimental Results," *Ocean Engineering*, Vol. **60** (2013), pp. 1–13.

12

Control of AUVs

12.1 Introduction

This chapter is devoted to the control of autonomous underwater vehicles (AUVs). Based on the dynamic models derived in Chapter 11, three AUV control problems are investigated including longitudinal gliding stabilization, yaw angle regulation, and spiral path tracking, using passivity-based control, sliding mode control, and two-degree-of-freedom control, respectively. Here, we would like to reiterate that a dynamic model of AUVs is a necessity when we consider the control of AUVs where a kinematic model is insufficient due to system nonlinearity and coupling effects between control inputs. The control problems are thus challenging due to the resulting increased system complexity compared to the kinematic models adopted for surface vehicles.

12.2 Longitudinal Gliding Stabilization

In this section, we look into the longitudinal plane stabilization problem of an underwater glider with an elevator-type control surface. The dynamics of an underwater glider in the longitudinal plane is first reviewed and separated into the slow dynamics and fast dynamics based on singular perturbation analysis. A passivity-based nonlinear controller for the approximated reduced model is proposed. Simulation results are then presented to show the effectiveness of the designed controller.

Passivity-based nonlinear control is an approach that stabilizes dynamical systems by creating passive systems via feedback control. A passive system consumes or dissipates energy rather than producing energy. A system is passive if the power inflow of the system, represented by the product of system input u and output y, is greater than or equal to the changing rate of the system stored

Mobile Robots: Navigation, Control and Sensing, Surface Robots and AUVs,
Second Edition. Gerald Cook and Feitian Zhang.
© 2020 by The Institute of Electrical and Electronics Engineers, Inc.
Published 2020 by John Wiley & Sons, Inc.

energy, represented by the time derivative of a positive semidefinite storage function $\dot{V}$. For example, an RLC circuit consisting of resisters, capacitors, and inductors is a passive system if we consider the input voltage for the RLC circuit as the system input, and the total current of the circuit as the system output. Using V as the Lyapunov function, we can design a feedback controller $u = -\phi(y)$ such that $\dot{V} \le -y\phi(y) \le 0$. The system origin is then stabilized by the LaSalle's invariance principle.

12.2.1 Longitudinal Dynamic Model Reduction

Review of the Longitudinal Model First, let us briefly review the longitudinal dynamic model of an underwater glider introduced in Section 11.5. We will then rewrite the dynamic model in another set of state variables for the convenience of control design. Using the same definitions, we have three coordinate reference frames (Figure 12.1) for describing the system dynamics including the body-fixed-reference frame $Ox_by_bz_b$, the velocity reference frame $Ox_vy_vz_v$, and the inertia reference frame $Axyz$. As shown in Figure 12.1, the external forces acting on the underwater glider include gravitational force, buoyancy force, lift force L, drag force D, and control force F_δ. The gravitational force and buoyancy force are in the opposite directions and the difference is described by the net buoyancy force m_0g. The control force F_δ, acting in Oz_b direction, is a hydrodynamic force coming from the control surface (elevator) traveling through the fluid medium. The control surface angle δ is defined as the angle between the control surface plane and the Ox_by_b plane. The lift force L and drag force D are another two hydrodynamic forces due to the relative motion between the vehicle and the fluid.

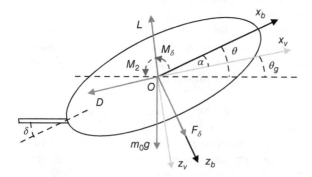

Figure 12.1 The schematic of an underwater glider with forces and moments defined in the corresponding coordinate frames (side view).

Simplified from the hydrodynamic model in Section 11.4, the hydrodynamic forces that are dependent on the angle of attack α and the velocity V can be described as follows:

$$L = (K_{L0} + K_L\alpha)V^2 \tag{12.1}$$

$$D = (K_{D0} + K_D\alpha^2)V^2 \tag{12.2}$$

$$F_\delta = K_{F_\delta}V^2 u_\delta \tag{12.3}$$

where K_{L0}, K_L are lift coefficients, and K_{D0}, K_D are drag coefficients. u_δ is the effective angle of attack that the control surface contributes to the underwater glider. There is a linear relationship between u_δ and the control surface angle δ; $u_\delta = K_{u_\delta}\delta$, where K_{u_δ} is a scale constant. K_{F_δ} is the coupling factor that describes the additional force that the control surface induces.

There are two moments about the Oy_b axis, which rotate the vehicle to a specific attitude. One is the hydrodynamic pitch moment M_2, and the other is the control moment M_δ. They are modeled as

$$M_2 = (K_{M0} + K_M\alpha + K_{q2}\omega_2)V^2 \tag{12.4}$$

$$M_\delta = -K_M u_\delta V^2 \tag{12.5}$$

where K_{M0} and K_M are pitch moment coefficients, K_q is the pitching damping coefficient, and ω_2 is the angular velocity for the pitch.

We take the assumptions (Bhatta and Leonard, 2008) that the movable mass is fixed at the origin O (during steady gliding), with the stationary mass distributed uniformly, and the added masses are equally valued in Ox_b and Oz_b directions. Rewriting the dynamic model (11.15)–(11.20) by applying state transformation, the underwater glider dynamics are obtained as

$$\dot{V} = -\frac{1}{m}\left(m_0 g \sin\theta_g + D - F_\delta \sin\alpha\right) \tag{12.6}$$

$$\dot{\theta}_g = \frac{1}{mV}\left(-m_0 g \cos\theta_g + L + F_\delta \cos\alpha\right) \tag{12.7}$$

$$\dot{\alpha} = \omega_2 - \frac{1}{mV}\left(-m_0 g \cos\theta_g + L + F_\delta \cos\alpha\right) \tag{12.8}$$

$$\dot{\omega}_2 = \frac{1}{J_2}\left(K_{M0} + K_M\alpha + K_{q2}\omega_2 - K_M u_\delta\right)V^2 \tag{12.9}$$

where m is the sum of the mass of the underwater glider and the added mass in Ox_b direction, J_2 is the total inertia about Oy_b axis, consisting of stationary mass inertia and added inertia in water, the gliding angle $\theta_g = \theta - \alpha$ where θ is the pitch angle, and g represents the gravitational acceleration.

For the open-loop system (i.e. $u_\delta = 0$), the steady gliding profile can be obtained from (12.6) to (12.9). The state variables at the equilibrium have the following relationships

$$\theta_{g_e} = \arctan \frac{-K_{D_e}}{K_{L_e}}, \qquad \alpha_e = -\frac{K_{M0}}{K_M}, \qquad \omega_{2e} = 0, \qquad V_e = \left(\frac{|m_0|g}{\sqrt{K_{D_e}^2 + K_{L_e}^2}} \right)^{\frac{1}{2}}$$

where $K_{D_e} = K_{D0} + K_D \alpha_e^2, K_{L_e} = K_{L0} + K_L \alpha_e$.

System Reduction via Singular Perturbation Bhatta and Leonard (2008) have shown with singular perturbation analysis that for the above open-loop system, the dynamic model can be reduced to a second-order system with good approximation, and the corresponding nondimensional full state model is:

$$\frac{d\bar{V}}{dt_n} = -\frac{m_0 g \sin\left(\bar{\theta}_g + \theta_{g_e}\right) + D - F_\delta \sin\left(\bar{\alpha} + \alpha_e\right)}{K_{D_e} V_e^2} \tag{12.10}$$

$$\frac{d\bar{\theta}_g}{dt_n} = -\frac{m_0 g \cos\left(\bar{\theta}_g + \theta_{g_e}\right) - L - F_\delta \cos\left(\bar{\alpha} + \alpha_e\right)}{K_{D_e} V_e^2 (1 + \bar{V})} \tag{12.11}$$

$$\varepsilon_1 \frac{d\bar{\alpha}}{dt_n} = \bar{\omega}_2 + \varepsilon_1 \frac{m_0 g \cos\left(\bar{\theta}_g + \theta_{g_e}\right) - L - F_\delta \cos\left(\bar{\alpha} + \alpha_e\right)}{K_{D_e} V_e^2 (1 + \bar{V})} \tag{12.12}$$

$$\varepsilon_2 \frac{d\bar{\omega}_2}{dt_n} = -(\bar{\alpha} + \bar{\omega}_2 - u_\delta)(1 + \bar{V})^2 \tag{12.13}$$

where the new state variables are defined as

$$\bar{V} = \frac{V - V_e}{V_e}, \qquad \bar{\theta}_g = \theta_g - \theta_{g_e}, \qquad \bar{\alpha} = \alpha - \alpha_e, \qquad \bar{\omega}_2 = \frac{K_q}{K_M} \omega_2$$

the nondimensional time t_n and some related constants are defined as

$$\tau_s = \frac{m}{K_{De} V_e}, \qquad \varepsilon_2 = -\frac{J_2}{K_q V_e^2} \frac{1}{\tau_s}, \qquad t_n = t/\tau_s, \qquad \varepsilon_1 = \frac{K_q}{K_M} \frac{1}{\tau_s}$$

For the new state model, the hydrodynamic forces and moment can be described as

$$D = \left(K_{D0} + K_D(\bar{\alpha} + \alpha_e)^2\right) V_e^2 (1 + \bar{V})^2 \tag{12.14}$$

$$L = \left(K_{L0} + K_L(\bar{\alpha} + \alpha_e)\right) V_e^2 (1 + \bar{V})^2 \tag{12.15}$$

$$M_2 = \left(K_{M0} + K_M \alpha + K_q \bar{\omega}_2\right) V_e^2 (1 + \bar{V})^2 \tag{12.16}$$

$$F_\delta = K_{F_\delta} u_\delta V_e^2 (1 + \bar{V})^2 \tag{12.17}$$

The system can be further written in a compact form

$$\frac{d\xi}{dt_n} = f(\xi, \eta, u_\delta) \tag{12.18}$$

$$\mu \frac{d\eta}{dt_n} = Ag(\xi, \eta, \varepsilon, u_\delta) \qquad (12.19)$$

where

$$\xi = \begin{bmatrix} \bar{V} \\ \bar{\theta}_g \end{bmatrix}, \quad \eta = \begin{bmatrix} \bar{\alpha} \\ \bar{\omega}_2 \end{bmatrix}, \quad f = \begin{bmatrix} f_1 \\ f_2 \end{bmatrix}, \quad g = \begin{bmatrix} g_1 \\ g_2 \end{bmatrix}$$

$$A = \begin{bmatrix} \dfrac{\mu}{\varepsilon_1} & 0 \\ 0 & \dfrac{\mu}{\varepsilon_2} \end{bmatrix}, \quad \varepsilon = \begin{bmatrix} \varepsilon_1 \\ \varepsilon_2 \end{bmatrix}, \quad \mu = \max(\varepsilon_1, \varepsilon_2)$$

and f and g are defined accordingly based on (12.10)–(12.13).

From singular perturbation analysis, by taking the limit of μ to zero, we arrive at $\bar{\omega}_2 = 0$ and $\bar{\alpha} = u_\delta$. Plugging those two fast-mode states into the other two state equations, the reduced model for the full system is obtained. Now we further set $\bar{\alpha} = 0$ in the reduced model for design convenience, since $\bar{\alpha}$ is relatively small in value. Then the approximation of the reduced model can be expressed as

$$\frac{d\xi}{dt_n} = f(\xi, 0, u_\delta) \qquad (12.20)$$

We will use this second-order system for the controller design with the expectation that the controller design based on the approximated reduced system will work for the original full system sufficiently well.

12.2.2 Passivity-Based Controller Design

The control objective is to design a feedback controller to stabilize the origin of the approximated reduced model. The open-loop reduced model (12.20) with $u_\delta = 0$ has an exponentially stable equilibrium point at the origin, which can be proven by Lyapunov analysis with the following positive definite Lyapunov function (Bhatta and Leonard, 2008)

$$\Phi = \frac{2}{3} - (1 + \bar{V})\cos\bar{\theta}_g + \frac{1}{3}(1 + \bar{V})^3 \qquad (12.21)$$

and $\dfrac{\partial \Phi}{\partial \xi} f(\xi, 0, 0) \le -b_1 \|\xi\|$ with $b_1 > 0$.

The approximated reduced system is nonlinear in dynamics but linear in control

$$\frac{d\xi}{dt_n} = f(\xi, 0, 0) + g_r(\xi)u_\delta \qquad (12.22)$$

where

$$g_r(\xi) = \begin{bmatrix} K_{F_\delta}(1 + \bar{V})^2 \sin \alpha_e / K_{D_e} \\ K_{F_\delta}(1 + \bar{V}) \cos \alpha_e / K_{D_e} \end{bmatrix} \tag{12.23}$$

We choose to design a nonlinear controller using passivity theory considering the nonlinear dynamics, the availability of a Lyapunov function, and the expectation for faster convergence. For passivity-based controller design, an output y_r needs to be defined for the approximated reduced system, to make the system passive (Khalil, 2002). The output is chosen as

$$y_r = \frac{\partial \Phi}{\partial \xi} g_r(\xi) \tag{12.24}$$

where

$$\frac{\partial \Phi}{\partial \xi} = \begin{bmatrix} \frac{\partial \Phi}{\partial \bar{V}} & \frac{\partial \Phi}{\partial \bar{\theta}_g} \end{bmatrix} = \begin{bmatrix} -\cos \bar{\theta}_g + (1 + \bar{V})^2 & (1 + \bar{V}) \sin \bar{\theta}_g \end{bmatrix} \tag{12.25}$$

We check the following expression for the approximated reduced model

$$\frac{d\Phi}{dt_n} = \frac{\partial \Phi}{\partial \xi} (f(\xi, 0, 0) + g_r(\xi) u_\delta)$$

Knowing that $\frac{\partial \Phi}{\partial \xi} f(\xi, 0, 0) \le 0$, we have

$$\frac{d\Phi}{dt_n} \le u_\delta y_r$$

Then by the definition of a passive system, the following system

$$\begin{cases} \dfrac{d\xi}{dt_n} = f(\xi, 0, u_\delta) \\ y_r = \dfrac{\partial \Phi}{\partial \xi} g_r(\xi) \end{cases} \tag{12.26}$$

is passive. Let control u_δ for system (12.22) be

$$u_\delta = -\phi(y_r) \tag{12.27}$$

for some function ϕ, where $y_r u_\delta = -y_r \phi(y_r) \le 0$.

Now we take Φ in (12.21) as the Lyapunov function for the closed-loop system (12.26). Then

$$\frac{d\Phi}{dt_n} = \frac{\partial \Phi}{\partial \xi} (f(\xi, 0, 0) + g_r(\xi) u_\delta) = \frac{\partial \Phi}{\partial \xi} f(\xi, 0, 0) + \frac{\partial \Phi}{\partial \xi} g_r(\xi) u_\delta \le -b_1 \|\xi\| + y_r u_\delta.$$

For AUV control, there is limitation on the magnitude of the control variable u_δ, so in this chapter, we take

$$\phi(y_r) = \frac{1}{K_c}\arctan(y_r) \qquad (12.28)$$

where K_c is the control parameter that is used to limit the control output magnitude. We then have

$$\frac{d\Phi}{dt_n} \leq -b_1\|\xi\| - \frac{1}{K_c}y_r\arctan(y_r) \qquad (12.29)$$

which proves the asymptotic stability of the origin. Furthermore, the additional negative term $-\dfrac{1}{K_c}y_r\arctan(y_r)$ in the derivative of Lyapunov function provides an extra stabilization advantage. With that term, the Lyapunov function will converge to zero more quickly, which results in a faster convergence speed. That would help the underwater glider to return to its steady gliding path with less time. We also want to point out that while the designed passivity-based controller guarantees the global stability of the origin in the reduced-order model (12.20), it is challenging to establish the global stability of the full-order closed-loop longitudinal dynamic model (12.18) and (12.19). Alternatively, the local stability can be established by linearization of the full system and checking the Hurwitz property of the linearized system matrix (Zhang and Tan, 2015). In addition, we conjecture that this controller will be similarly effective for the full system as it does for the approximated reduced system. In particular, we anticipate that the controller will provide a faster convergence speed than the open-loop controller $u_\delta = 0$, due to the additional negative term it introduced into (12.29).

12.2.3 Simulation Results

To evaluate the control performance, we simulate the full dynamic model using the designed passivity-based controller. The underwater glider parameters we used in simulation are: $m = 10$ kg, $J_2 = 0.08$ kg m^2, $K_{L0} = 0$ kg/m, $K_L = 303.6$ kg/m, $K_{D0} = 3.15$ kg/m, $K_D = 282.8$ kg/m, $K_q = -0.8$ kg, $K_{M0} = 0.39$ kg, $K_M = -14.7$ kg, $K_{u_\delta} = 29.5$, and $m_0 = 0.05$ kg. The equilibrium point is $V_e = 0.24$ m/s, $\theta_{g_e} = -22.5°$, $\alpha_e = 1.52°$, and $\omega_{2e} = 0$ rad/s.

Suppose a current disturbance makes the vehicle deviate from its steady gliding path. The control objective here is to stabilize the dynamic system back to its equilibrium point, thus driving the vehicle back to its steady gliding status. The initial states are given as $V_0 = 0.2$ m/s, $\theta_{g_0} = -35°$, $\alpha_0 = 1°$, and $\omega_{20} = 0$ rad/s. In simulation, we also consider the dynamics of the actuator for moving the control surface, approximated by a first-order system with a time constant of 10 ms. The simulation time is 60 seconds.

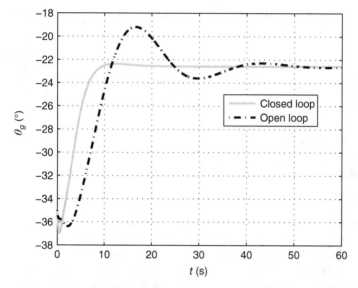

Figure 12.2 Simulation results on the trajectories of the gliding angle θ_g for the open-loop $u_\delta = 0$ and closed-loop ($K_c = 2$) cases, respectively.

Figure 12.2 shows that the passivity-based controller designed for the reduced model works for the original full-order system, not only stabilizing the steady gliding equilibrium but also speeding up the convergence process as we expected from the analysis. Figures 12.3 and 12.4 show the influences of the control parameter K_c on the control output and the glide angle transients. With a smaller K_c, the system converges faster but requires larger initial control effort. In addition, using the arctangent function for the tunable parameter K_c as in (12.28) makes it very convenient to balance between the control effort and the convergence speed.

12.3 Yaw Angle Regulation

12.3.1 Problem Statement

Steady gliding motion is the most commonly used profile for underwater gliders, providing the capability of sampling water in the field while saving energy at the same time. Setting the right-hand side of the dynamic equations of underwater gliders (11.2)–(11.5) to zero, one can solve those equations for the steady glide path given a fixed movable mass displacement r_p and excess mass m_0, with zero control surface (e.g. rudder) angle (Zhang et al., 2013). Due to the existence of ambient currents or disturbances, underwater vehicles are susceptible to yaw

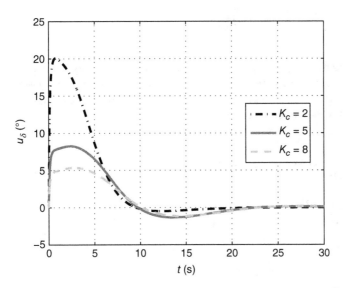

Figure 12.3 Plot of control u_δ for the closed-loop simulation with different values for K_c.

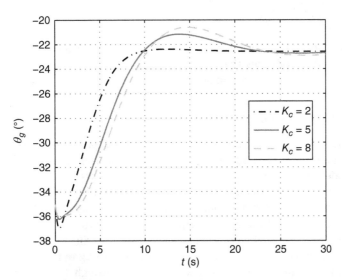

Figure 12.4 Plot of gliding angle θ_g for the closed-loop simulation with different values for K_c.

deviation from its desired direction, beside the longitudinal-plane perturbation discussed in Section 12.2, which makes yaw angle stabilization critical.

For succinctness, we first rewrite the system dynamics equations (11.2)–(11.5) in a compact form

$$\dot{x} = f(x) + \Delta_1(x) + g(x)(u + \Delta_2(t, x, u)) \tag{12.30}$$

$$y = h(x) = \psi \tag{12.31}$$

where x is the system state, $x = [\phi\ \theta\ \psi\ v_1\ v_2\ v_3\ \omega_1\ \omega_2\ \omega_3]^T$, $u = \delta$ is the rudder angle and the control input in the current setting, and $\Delta_1(x)$ and $\Delta_2(t, x, u)$ represent system uncertainties. The system output is chosen to be the yaw angle ψ. The function $g(x)$ is dependent on the state, i.e.

$$g(x) = \begin{bmatrix} 0_{3\times 1} \\ -\dfrac{1}{2m_1}\rho V^2 SC_{SF}^\delta \cos\alpha \sin\beta \\ \dfrac{1}{2m_2}\rho V^2 SC_{SF}^\delta \cos\beta \\ -\dfrac{1}{2m_3}\rho V^2 SC_{SF}^\delta \sin\alpha \sin\beta \\ -\dfrac{1}{2J_1}\rho V^2 SC_{M_Y}^\delta \sin\alpha \\ 0 \\ \dfrac{1}{2J_3}\rho V^2 SC_{M_Y}^\delta \cos\alpha \end{bmatrix} \tag{12.32}$$

The yaw angle stabilization problem in this chapter is to design a state-feedback controller that stabilizes the system and regulates the yaw angle ψ to a desired value r using the rudder angle δ in the presence of disturbances.

12.3.2 Sliding Mode Controller Design

Sliding model control is a practical nonlinear control method, particularly useful for robust stabilization of nonlinear systems with uncertainties (Khalil, 2002). Sliding mode control is a means of forcing a portion of a systems dynamics to behave according to a specified time constant. This can be achieved as long as the required control effort does not exceed the actuator capability. Suppose the desired behavior is for the system output to recover from disturbances according to a time constant τ. Then one constructs a switching line in the output-rate (time derivative of output) versus output plane according to the equation, $\tau \times$ output rate $=$ $-$output. Clearly this line passes through the origin. Whenever the point based on actual output and actual output rate is below this line, the control signal is set to positive (limit). When the point just described is above this line, the control signal is set to negative (limit). When operating within its allowed limits, the system trajectory is driven to the switching line and then moves along this line toward the origin, slightly crossing back and forth across the line as the control chatters between its positive and negative values. The net result is the prescribed dynamic behavior. We apply sliding mode control theory to solve the yaw angle regulation problem for underwater gliders. Control design follows the procedures described in Khalil (2002). Furthermore, to facilitate implementation, we propose a simplified sliding mode controller that requires only partial state feedback.

In order to obtain the relative degree of the system, we take the time derivatives of $h(x)$

$$\dot{h}(x) = \dot{\psi} = \sin\phi\sec\theta\omega_2 + \cos\phi\sec\theta\omega_3 = L_f h(x) \tag{12.33}$$

$$\ddot{h}(x) = \ddot{\psi} = L_f^2 h(x) + L_{\Delta_1} L_f h(x) + L_g L_f h(x)(u + \Delta_2(t,x,u)) \tag{12.34}$$

where $L_f h(\cdot)$ represents the Lie derivative of function $h(\cdot)$ with respect to the vector field $f(\cdot)$ (Khalil, 2002), and $L_f^2 h(x)$ is equal to $L_f L_f h(x)$.

The fact that $\dot{h}(x)$ does not depend on control input u and $\ddot{h}(x)$ does, implies that

$$L_g h(x) = 0 \tag{12.35}$$

$$L_g L_f h(x) \neq 0 \tag{12.36}$$

Therefore, the relative degree of the system $\rho_{sys} = 2$.

From Frobenius Theorem (Warner, 1983), there exists a transform function $T(x)$, which converts the original system to the normal form with system states $[\eta\ \xi]^T$.

$$\begin{bmatrix}\eta\\\xi\end{bmatrix} = \begin{bmatrix}\Phi(x)\\\Psi(x)\end{bmatrix} = \begin{bmatrix}p_1(x)\\\vdots\\p_7(x)\\h(x)\\L_f h(x)\end{bmatrix} = T(x) \tag{12.37}$$

where

$$\frac{\partial p_i}{\partial x} g(x) = 0, \quad \text{for } i = 1,2,\ldots,7 \tag{12.38}$$

$$\dot{\xi}_1 = \xi_2 \tag{12.39}$$

Let r denote the reference trajectory for the yaw angle, which is a constant in yaw angle regulation. Take

$$\mathcal{R} = \begin{bmatrix}r\\\dot{r}\end{bmatrix} \tag{12.40}$$

The yaw error vector e is

$$e = \xi - \mathcal{R} = \begin{bmatrix} \xi_1 - r \\ \xi_2 - \dot{r} \end{bmatrix} \qquad (12.41)$$

Then the error dynamics is expressed as

$$\dot{\eta} = f_0(\eta, \xi) \qquad (12.42)$$

$$\dot{e}_1 = e_2 \qquad (12.43)$$

$$\dot{e}_2 = L_f^2 h(x) + L_{\Delta_1} L_f h(x) + L_g L_f h(x)(u + \Delta_2(t, x, u)) - \ddot{r} \qquad (12.44)$$

Assume that $\dot{\eta} = f_0(\eta, \xi)$ is bounded-input-bounded-state stable with ξ as the input. We design a sliding manifold

$$s = e_2 + k_0 e_1 \qquad (12.45)$$

where k_0 is a positive constant.

A sliding mode controller is designed to cancel the known terms as in feedback linearization, i.e.,

$$u = -\frac{1}{L_g L_f h(x)}\left(k_0 e_2 + L_f^2 h(x) - \ddot{r}\right) + v \qquad (12.46)$$

where v is the switching component, and

$$L_f^2 h(x) = \frac{\partial \dot{\psi}}{\partial x} f \qquad (12.47)$$

$$L_g L_f h(x) = \frac{\partial \dot{\psi}}{\partial x} g \qquad (12.48)$$

$$\frac{\partial \dot{\psi}}{\partial x} = \begin{bmatrix} \cos\phi \sec\theta \omega_2 \\ \sec\theta \tan\theta(\cos\phi\omega_2 + \cos\phi\omega_3) \\ 0_5 \times 1 \\ \sin\phi \sec\theta \\ \cos\phi \sec\theta \end{bmatrix}^T \qquad (12.49)$$

Or we can take the controller as the pure switching component

$$u = v \qquad (12.50)$$

Then in either case, the $\dot{s}$-equation can be written as

$$\dot{s} = L_g L_f h(x)v + \Delta(t, x, v) \qquad (12.51)$$

Suppose that the uncertainty satisfies the following inequality,

$$\left|\frac{\Delta(t,x,\nu)}{L_g L_f h(x)}\right| \leq \rho(x) + \kappa_0 |\nu|, \quad 0 \leq \kappa_0 \leq 1 \tag{12.52}$$

where $\rho(x)$ represents the upper bound of the uncertainty related to the system states.

Design the switching component

$$\nu = -\gamma(x)\text{sat}(s/\epsilon) \tag{12.53}$$

where $\text{sat}(s/\epsilon)$ is a high-slope saturation function with a small constant ϵ, used to reduce chattering, and $\gamma(x) \geq \rho(x)/(1 - \kappa_0) + \gamma_0$ with constant γ_0 to deal with the nonvanishing disturbance $\Delta(t, x, \nu)$ if that is the case.

In this chapter, we choose

$$\gamma(x) = k_1 \|x - x_e\|_2^{k_2} + k_3 \tag{12.54}$$

where x_e is the system equilibrium point, calculated given a steady gliding profile as in Section 11.6, and k_1, k_2, k_3 are controller parameters for tuning closed-loop dynamic system performances, partially determined by the uncertainty type.

Based on the fact that the yaw angle ψ is the state we care about, we further simplify the sliding mode controller (equations 12.50, 12.53, and 12.54) to

$$u = -\left(k_1 \|\psi - \psi_e\|_2^{k_2} + k_3\right)\text{sat}(s/\epsilon) \tag{12.55}$$

where the sliding mode controller only requires the feedback information of the yaw angle that is typically available for AUVs. The effect of this simplification on stability can be compensated by increasing controller parameters k_1, k_2 and especially the nonzero constant k_3. Although this simplification will increase the tracking error in general, the controller implementation becomes much simpler.

12.3.3 Simulation Results

To evaluate the designed sliding mode controller, simulation is carried out. The parameters used in the simulation are shown in Table 12.1.

The initial state values for the simulation are

$$\phi = 0\,\theta = -30°, \psi = 30°, \nu_1 = 0.27\ \text{m/s}, \nu_2 = 0, \nu_3 = 0, \omega_1 = 0, \omega_2 = 0, \omega_3 = 0$$

The controller parameters used in the simulation are

$$k_0 = 10, k_1 = 10/30/50, k_2 = 0.8/1/1.2\ k_3 = 0.01$$

Figures 12.5–12.7 show the trajectories of the yaw angle ψ, the controller command δ, and the sideslip angle β when varying the controller parameter

Table 12.1 System parameters used in simulation.

Parameter	Value	Parameter	Value
m_1	8.0 kg	m_2	19.8 kg
m_3	10.8 kg	$\bar{m}$	1.6 kg
C_{D0}	0.45	C_D^α	17.59 rad^{-2}
$C_{F_S}^\beta$	−2 rad^{-1}	$C_{F_S}^\delta$	1.5 rad^{-1}
C_{L0}	0.075	C_L^α	19.58 rad^{-1}
J_1	1.27 kg·m^2	J_2	0.08 kg·m^2
J_3	0.13 kg·m^2	C_{M_0}	0.0076 m
$C_{M_R}^\beta$	−0.3 m/rad	$C_{M_P}^\alpha$	0.71 m/rad
$C_{M_Y}^\beta$	5 m/rad	$C_{M_Y}^\delta$	−0.2 m/rad
K_{q1}	−0.16 m·s/rad	K_{q2}	−0.80 m·s/rad
K_{q3}	−0.16 m·s/rad	S	0.019 m^2

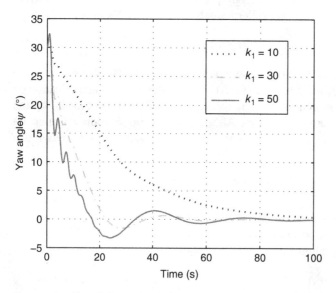

Figure 12.5 Plot of the yaw angle with respect to different controller parameters k_1.

k_1, and Figures 12.8–12.10 show the simulation results when varying controller parameter k_2. From the results, we observe that under proper controller parameters, the sliding mode controller is able to regulate the yaw angle, which is deviated from the desired orientation, back to the original, zero angle within

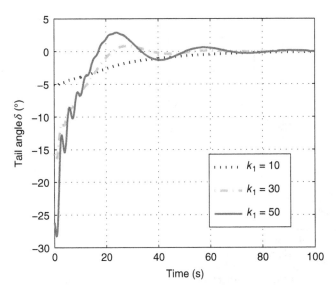

Figure 12.6 Plot of the rudder angle with respect to different controller parameters k_1.

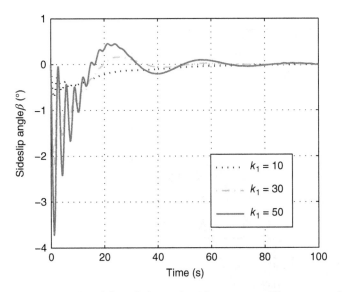

Figure 12.7 Plot of the sideslip angle with respect to different controller parameters k_1.

a relative short time. Consequently, the trajectory of the glider is adjusted to the desired path, with the heading orientation being zero degree, as shown in Figure 12.11. From the comparison under different controller parameters, we find that k_1 and k_2 control the balance between response speed and control

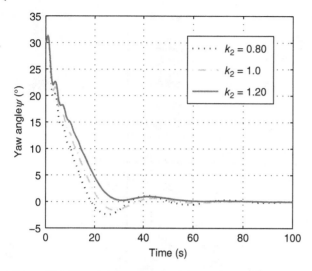

Figure 12.8 Plot of the yaw angle with respect to different controller parameters k_2.

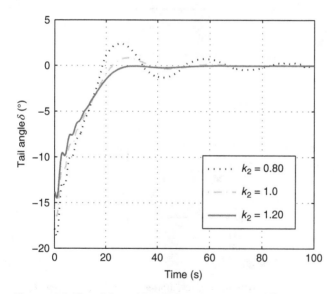

Figure 12.9 Plot of the rudder angle with respect to different controller parameters k_2.

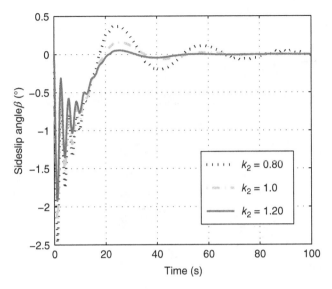

Figure 12.10 Plot of the sideslip angle with respect to different controller parameters k_2.

effort. With larger k_1 and smaller k_2, the system responses faster and requires a control output with a larger amplitude. Regarding parameter k_3, as in the sliding mode design principle, it should balance the steady-state error and uncertainty tolerance capability. With a larger k_3, the controller is able to work under larger uncertainty while leading to bigger steady-state error.

12.4 Spiral Path Tracking

12.4.1 Steady Spiral and Its Differential Geometric Parameters

The three-dimensional path tracking of underwater gliders is very challenging because the influences of the control inputs on the vehicle's locomotion are strongly nonlinear and coupled. It is more convenient to look into the influence of control inputs on the vehicle's differential geometry features, such as curvature and torsion, because we can examine the relationship between those geometric characteristic parameters and the control inputs by studying the steady-state spiral motions.

We decompose an arbitrary three-dimensional curve into a set of continuously evolving spirals. In this way, at any point of the space curve, there is an imaginary matching spiral path with the same curvature and torsion. With this interpretation, instead of using the Euclidean positions, we will explore the task of three-dimensional path tracking via designing and following continuously

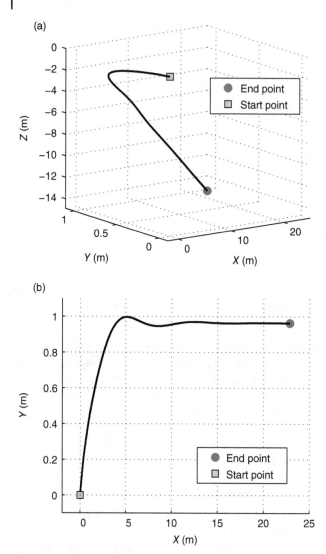

Figure 12.11 Trajectory of the underwater glider for controller parameters $k_1 = 30$, $k_2 = 1$. (a) Three-dimensional view; (b) top view for X–Y plane.

evolving spirals from the point of view of differential geometry (Zhang and Tan, 2014).

First, let us review the results of the steady spiral motion discussed in Section 11.8. There are three control variables available to manipulate the vehicle's motion profile: the excess mass m_0, the position of the movable mass r_p, and the rudder angle δ. When all three controls are fixed at nonzero values, underwater gliders will perform three-dimensional spiraling motion and finally enter a steady spiral, where the yaw angle ψ changes at a constant rate while the roll angle ϕ and pitch angle θ remain constant.

The dynamics of the spiral motion, derived from (11.2) to (11.5), can be presented in a compact form as

$$\dot{x}_s = f(x_s, u) = [f_i(x_s, u)]_{8 \times 1} \tag{12.56}$$

where system states $x_s = [\phi \ \theta \ v_1 \ v_2 \ v_3 \ \omega_1 \ \omega_2 \ \omega_3]^T$, and control inputs $u = [r_p \ m_0 \ \delta]^T$.

The steady-state spiraling equations are obtained by setting time derivatives to zero in (12.56)

$$0 = f(x_s, u) = [f_i(x_s, u)]_{8 \times 1} \tag{12.57}$$

In a steady spiral, $R^T k$ is constant since

$$R^T k = R^T \begin{pmatrix} 0 \\ 0 \\ 1 \end{pmatrix} = \begin{pmatrix} -\sin\theta \\ \sin\phi\cos\theta \\ \cos\phi\cos\theta \end{pmatrix} \tag{12.58}$$

The angular velocity has only one degree of freedom with ω_{3i} in Ox axis in the inertial frame

$$\omega_b = \omega_{3i} \left(R^T k \right) \tag{12.59}$$

Therefore, in the system of algebraic equation (12.57), there are nine independent variables (including control inputs) for describing the steady spiral motion: $[\phi \ \theta \ \omega_{3i} \ r_p \ m_0 \ \delta \ V \ \alpha \ \beta]^T$. Hereafter, we will use a state transformation on linear velocity variables for the sake of calculation convenience

$$v_b = \begin{pmatrix} v_1 \\ v_2 \\ v_3 \end{pmatrix} = R_{bv} \begin{pmatrix} V \\ 0 \\ 0 \end{pmatrix} = \begin{pmatrix} V\cos\alpha\cos\beta \\ V\sin\beta \\ V\sin\alpha\cos\beta \end{pmatrix} \tag{12.60}$$

In the elementary differential geometry, a three-dimensional curve is captured by its curvature and torsion. The curvature κ is the amount by which a geometric object deviates from being flat, or the degree by which a geometric object bends, while torsion τ measures the departure of a curve from a plane, or how sharply a curve twists. Any time-trajectory of a smooth space curve can be completely described mathematically using curvature, torsion, and velocity with Frenet–Serret formulas (Pressley, 2010).

The geometric parameters (curvature, torsion, velocity) of a steady spiral can be expressed as

$$\kappa = \frac{r}{r^2 + c^2} \tag{12.61}$$

$$\tau = \frac{c}{r^2 + c^2} \tag{12.62}$$

where r is the steady spiral radius, and $2\pi c$ is the steady spiral pitch, or the vertical separation between two steady spirals. Furthermore,

$$r = \frac{V_h}{\omega_{3i}} \tag{12.63}$$

$$c = \frac{V_v}{\omega_{3i}} \tag{12.64}$$

where V_h and V_v are the horizontal velocity and vertical velocity, respectively, of the steady spiral motion.

We also have

$$V_h^2 + V_v^2 = V^2 \tag{12.65}$$

From (12.61) to (12.65), the angular velocity ω_{3i} and the vertical velocity V_v can be described by the three geometric parameters, κ, τ, and V

$$\omega_{3i} = V\sqrt{\kappa^2 + \tau^2} \tag{12.66}$$

$$V_v = V\sqrt{\frac{\tau^2}{\tau^2 + \kappa^2}} \tag{12.67}$$

12.4.2 Two Degree-of-Freedom Control Design

In this section, we propose a 2-DOF control strategy for the path tracking problem. Inverse mapping of steady spirals and robust H_∞ control for a linearized model are used as feedforward and feedback controllers, respectively. The feedforward controller serves as a driving force stabilizing the glider at desired steady spirals. The feedback H_∞ controller speeds up the convergence and enhances the system robustness. The idea of using a 2-DOF controller is that with the feedforward inverse mapping, the dynamic nonlinearity is reduced so that a feedback H_∞ controller can be designed based on the linearized model to achieve improved transient performances.

Feedforward Control via Inverse Mapping of Steady Spiral Motion Based on the fact that a three-dimensional path can be decomposed into a set of continuously evolving spirals, we propose a 3D path tracking approach that follows geometric characteristics of these spirals instead of Euclidean positions. We calculate the desired control inputs given a steady spiral that is parameterized by curvature, torsion, and velocity. This inverse mapping solution is used as an open-loop feedforward controller for the 3D path tracking problem.

From (12.59), it can be shown that the first two equations in (12.57) always hold, thus redundant. With the value of V_v known from (12.67), we have one more constraint equation

$$V_v = R_{bv} \begin{pmatrix} V \\ 0 \\ 0 \end{pmatrix} (R^T k) \tag{12.68}$$

Given κ, τ, and V, we calculate the value of the angular velocity ω_{3i} from (12.66). Knowing the values of V and ω_{3i}, there are seven unknown variables left out of nine independent states for the steady spiral motion $(\phi \ \theta \ \omega_{3i} \ r_p \ m_0 \ \delta \ V \alpha \beta)^T$. Correspondingly, there are seven independent algebraic equations from (12.57) and (12.67). The inverse mapping problem is then formulated as

$$0 = g(x) = [g_i(x)]_{7 \times 1} \tag{12.69}$$

where $x = (\phi \ \theta \ \alpha \ \beta \ r_p \ m_0 \ \delta)^T$. The expansion of $g_i(x)$ is shown in (12.70)–(12.76).

$$
\begin{aligned}
0 = {} & m_2 s\beta V c\phi c\theta \omega_{3i} - m_3 sac\beta V s\phi c\theta \omega_{3i} - m_0 gs\theta \\
& - 1/2 \rho V^2 S\left(C_{SF}^\beta \beta + C_{SF}^\delta \delta\right) cas\beta + 1/2\rho V^2 S\left(C_{L0} + C_L^\alpha \alpha\right) sa \\
& - 1/2 \rho V^2 S\left(C_{D0} + C_D^\alpha \alpha^2 + C_D^\delta \delta^2\right) cac\beta
\end{aligned} \tag{12.70}
$$

$$
\begin{aligned}
0 = {} & -m_3 sac\beta V s\theta \omega_{3i} - m_1 cac\beta V c\phi c\theta \omega_{3i} + m_0 gs\phi c\theta \\
& + 1/2 \rho V^2 S\left(C_{SF}^\beta \beta + C_{SF}^\delta \delta\right) c\beta - 1/2\rho V^2 S\left(C_{D0} + C_D^\alpha \alpha^2 + C_D^\delta \delta^2\right) s\beta
\end{aligned} \tag{12.71}
$$

$$
\begin{aligned}
0 = {} & m_1 cac\beta V s\phi c\theta \omega_{3i} + m_2 s\beta V s\theta \omega_{3i} + m_0 gc\phi c\theta \\
& - 1/2 \rho V^2 S\left(C_{SF}^\beta \beta + C_{SF}^\delta \delta\right) sas\beta - 1/2\rho V^2 S\left(C_{L0} + C_L^\alpha \alpha\right) ca \\
& - 1/2 \rho V^2 S\left(C_{D0} + C_D^\alpha \alpha^2 + C_D^\delta \delta^2\right) sac\beta
\end{aligned} \tag{12.72}
$$

$$
\begin{aligned}
0 = {} & (J_2 - J_3) s\phi c\theta c\phi c\theta \omega_{3i}^2 + (m_2 - m_3) s\beta sac\beta V^2 \\
& - 1/2 \rho V^2 S\left(C_{M_0} + C_{M_P}^\alpha \alpha + K_{q2} s\phi c\theta \omega_{3i}\right) cas\beta - m_w gr_w s\phi c\theta \\
& - 1/2 \rho V^2 S\left(C_{M_Y}^\beta \beta + K_{q3} c\phi c\theta \omega_{3i} + C_{M_Y}^\delta \delta\right) sa \\
& + 1/2 \rho V^2 S\left(C_{M_R}^\beta \beta - K_{q1} s\theta \omega_{3i}\right) cac\beta
\end{aligned} \tag{12.73}
$$

$$0 = (J_1 - J_3)s\theta c\phi c\theta \omega_{3i}^2 + (m_3 - m_1)ca c\beta sac\beta V^2 - m_w gr_w s\theta$$
$$- \bar{m}gr_p c\phi c\theta + 1/2\rho V^2 S\left(C_{M_R}^\beta \beta - K_{q1}s\theta\omega_{3i}\right)s\beta$$
$$+ 1/2\rho V^2 S\left(C_{M_0} + C_{M_P}^\alpha \alpha + K_{q2}s\phi c\theta\omega_{3i}\right)c\beta \tag{12.74}$$

$$0 = (J_2 - J_1)s\theta s\phi c\theta \omega_{3i}^2 + (m_1 - m_2)ca c\beta s\beta V^2$$
$$- 1/2\rho V^2 S\left(C_{M_0} + C_{M_P}^\alpha \alpha + K_{q2}s\phi c\theta\omega_{3i}\right)sas\beta + \bar{m}gr_p s\phi c\theta$$
$$+ 1/2\rho V^2 S\left(C_{M_Y}^\beta \beta + K_{q3}c\phi c\theta\omega_{3i} + C_{M_Y}^\delta \delta\right)ca$$
$$+ 1/2\rho V^2 S\left(C_{M_R}^\beta \beta - K_{q1}s\theta\omega_{3i}\right)sac\beta \tag{12.75}$$

$$0 = V_v/V + cac\beta s\theta - s\beta c\theta s\phi - sac\beta c\theta c\phi \tag{12.76}$$

Unfortunately, there is no closed-form solution to this system of equations. In this chapter, we use a Newton's method to find solutions recursively, which provides the desired open-loop control inputs. The iterative algorithm for Newton's method reads (Kelley, 2003)

$$\hat{x}_{i+1} = \hat{x}_i - J^{-1}(\hat{x}_i)g(\hat{x}_i) \tag{12.77}$$

Here $\hat{x}_i$ is the ith-step iteration for x, and $J(x)$ is the Jacobian matrix of $g(x)$

$$J(x,u) = \frac{\partial g}{\partial x} = \left(\frac{\partial g_i}{\partial x_j}\right)_{7\times 7} \tag{12.78}$$

2-DOF Control Design with a Feedback H_∞ Controller   Building upon the feedforward controller designed previously using inversing mapping of steady spirals, we propose a 2-DOF control strategy for the 3D path tracking problem, the configuration of which is shown in Figure 12.12. The transfer function $G(s)$

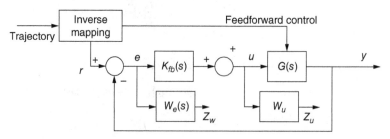

Figure 12.12 The control system diagram with a combination of open-loop control and closed-loop control.

represents the spiral dynamics system. $K_{fb}(s)$ is the feedback controller. Let $r = (\phi_r \, \theta_r \, v_{1r})^T$ be the reference signal calculated by the inverse mapping of given steady spirals described by curvature, torsion, and velocity, or the numerical solution of algebraic equations (12.70)–(12.76). Let $u = (r_p \, m_0 \, \delta)^T$ be the control input to the plant. $e = r - y$ represents the tracking error. $W_e(s)$ is the user-defined weighting function to impose the requirements for the tracking bandwidth and tracking error amplitude. The state-space realization of $W_e(s)$ is as follows,

$$\dot{x}_w = A_w x_w + B_w u_w \tag{12.79}$$

$$z_w = C_w x_w + D_w u_w \tag{12.80}$$

$W_u = \mathrm{diag}(w_{u1}, w_{u2}, w_{u3})$ is the weighting function to help balance the magnitude of system control inputs, and $z_u = W_u u$.

12.4.2.0.1 Linearized Model About a Steady Spiral Trajectory

We first derive a linearized model by linearization of system dynamics about the relative equilibrium of steady spiral trajectories. Recall the spiral dynamics (12.56)

$$\dot{x}_s = f(x_s, u) = [f_i(x_s, u)]_{8 \times 1}$$

where $x_s = (\phi \; \theta \; v_1 \; v_2 \; v_3 \; \omega_1 \; \omega_2 \; \omega_3)^T$, and $u = (r_p \; m_0 \; \delta)^T$. We define transformed system states $z = (\phi \; \theta \; V \; \alpha \; \beta \; \omega_1 \; \omega_2 \; \omega_3)^T$ for the convenience of computation of the Jacobian matrix $J(x_s, u)$. The linearized system matrices are

$$A = [J(x_s, u)]_e = \left[\frac{\partial f}{\partial x} \right]_e = \left[\frac{\partial f}{\partial z} \left(\frac{\partial x}{\partial z} \right)^{-1} \right]_e \tag{12.81}$$

$$B = \left[\frac{\partial f}{\partial u} \right]_e \tag{12.82}$$

Here $[\cdot]_e$ indicates the matrix elements are evaluated at the equilibrium point. We define the linearized system output as $y = (\phi \; \theta \; v_1)^T$, linear in system states. The linearized system output matrices are

$$C = [1_{3 \times 3} \quad 0_{3 \times 5}] \tag{12.83}$$

$$D = 0_{3 \times 3} \tag{12.84}$$

12.4.2.0.2 H_∞ Controller Design

The tracking performance can be characterized by the tracking error e. The control effort can be characterized by the control input u. The objective of the feedback control design is to minimize those two signals e and u. This optimization problem for the linearized model can be transformed into an H_∞ robust control framework as shown in Figure 12.13. In this H_∞ control system configuration,

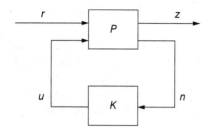

Figure 12.13 Transformed 2-DOF control configuration in H_∞ control framework.

$$K(s) = K_{fb}(s) \tag{12.85}$$

$$z = (z_w \quad z_u)^T \tag{12.86}$$

$$n = r - y \tag{12.87}$$

The interconnected system

$$P = \begin{pmatrix} A_p & B_p \\ C_p & D_p \end{pmatrix} \tag{12.88}$$

where

$$A_p = \begin{pmatrix} A & 0 \\ -B_wC & A_w \end{pmatrix} \quad B_p = \begin{pmatrix} 0 & B \\ B_w & -B_wD \end{pmatrix}$$

$$C_p = \begin{pmatrix} 0 & 0 \\ -D_wC & C_w \\ -C & 0 \end{pmatrix} \quad D_p = \begin{pmatrix} 0 & W_u \\ D_w & -D_wD \\ 1 & -D \end{pmatrix}$$

The design objective is then to minimize the H_∞ norm of the transfer function from **r** to **z**

$$\min_K \| T_r^z(s) \|_\infty \tag{12.89}$$

Here, $\| T_r^z(s) \|_\infty$ equals the maximum singular value of the transfer function $T_r^z(s)$ over the frequency domain.

To solve the above H_∞ optimization problem, we adopt the command *hinfmix* (·) in Matlab LMI toolbox. The output provides the feedback controller $K(s)$.

12.4.3 Simulation Results

Given desired trajectories of curvature, torsion, and velocity, simulation is conducted to test the effectiveness of the proposed 2-DOF control algorithm. For the purpose of comparison, we have also conducted simulation with pure inverse mapping control (see Section 12.4.2) and Proportional-Integral (PI) control. The PI controller is designed as

$$\delta = K_P^\delta \Delta \kappa + K_I^\delta \int \Delta \kappa \tag{12.90}$$

$$r_p = K_P^{r_p} \Delta \tau + K_I^{r_p} \int \Delta \tau \tag{12.91}$$

$$m_0 = K_P^{m_0} \Delta V + K_I^{m_0} \int \Delta V \tag{12.92}$$

where Δ stands for the difference between the desired value and actual value of the variable that follows. The particular form of the PI controller, where one control input is only dependent on the error feedback from one geometric parameter, is adopted for design convenience and based on the observed influences of the control inputs on the geometric parameters (Zhang et al., 2017), where it appears that each control input has more pronounced impact on one of the geometric parameters than other two inputs. The PI controller coefficients are designed as $K_P^\delta = 0.1, K_I^\delta = 0.01, K_P^{r_p} = 0.05, K_I^{r_p} = 0.005, K_P^{m_0} = 0.1$, and $K_I^{m_0} = 0.01$. There are three control inputs and three geometric variables to track, so the strong coupling between the control inputs makes the parameter tuning quite challenging. The PI control parameters are tuned in simulation in order to obtain the best tracking performance.

The model parameters used in simulation are the same as in Table 12.1. The initial values of system states used in simulation are

$$\theta = -7.2°, \phi = 0, v_1 = 0.1 \text{ m/s}, v_2 = 0, v_3 = 0.04°, \omega_1 = 0, \omega_2 = 0, \omega_3 = 0$$

This represents a longitudinal plane glide motion.

The weighting functions for the feedback H_∞ control design are chosen as

$$A_w = \text{diag}(-500, -100, -200) \quad B_w = \text{diag}(9,9,9)$$
$$C_w = \text{diag}(-1, -1, -1) \quad D_w = \text{diag}(0.1, 0.1, 0.1)$$
$$W_u = \text{diag}(0.2, 0.05, 0.2)$$

The solution to the H_∞ optimization problem is an eleventh-order linear system for the controller $K(s)$. Through model reduction techniques by investigating the dominant singular values, a seventh-order controller is used in simulation. Saturation is also imposed to restrict the control inputs to the feasible range of actuators.

Figures 12.14 and 12.15 show the simulated trajectories of the three geometric parameters and the three control inputs, respectively, for tracking a steady spiral path with constant geometric parameters. Figure 12.16 shows the tracking trajectory in 3D view under the proposed 2-DOF controller. From the simulation results, we see that both PI controller and feedforward inverse mapping

(a)

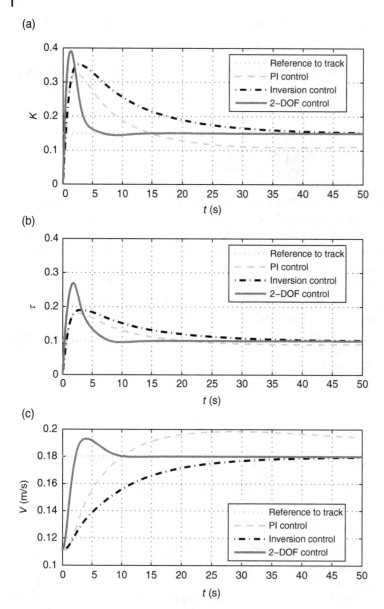

(b)

(c)

Figure 12.14 The simulation results of the geometric parameters when tracking a steady spiral trajectory. (a) Curvature; (b) torsion; (c) velocity.

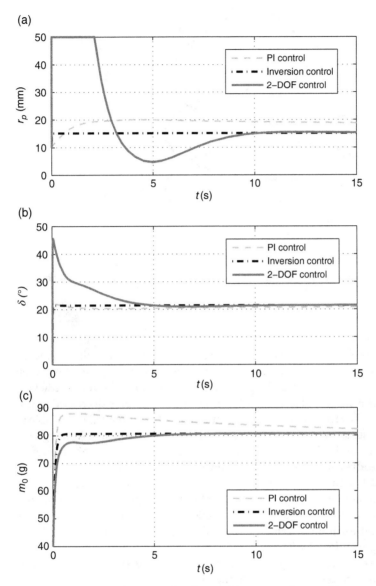

Figure 12.15 The simulation results of control inputs on when tracking a steady spiral trajectory. (a) Displacement of movable mass; (b) rudder angle; (c) net buoyancy.

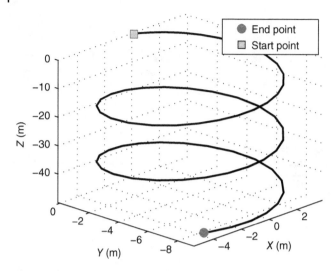

Figure 12.16 3D trajectory when tracking a steady spiral under the 2-DOF controller.

controller are able to stabilize the system to the desired steady-spiral trajectory. However, the feedforward controller results in convergence time between 30 and 40 seconds while the PI controller leads to a convergence time even greater than 50 seconds. With the proposed 2-DOF control, the system tracking performance is improved significantly with convergence time decreased to less than 10 seconds. Here we want to point out that all three controllers are able to achieve zero steady-state tracking error for fixed reference inputs. Although the tracking error under PI control is relatively large in the figures, given enough time, the steady-state error will eventually converge to zero.

Figures 12.17 and 12.18 show the simulated trajectories of the three reference geometric parameters and the three control inputs, respectively, when the reference velocity changes as a sinusoid function with respect to time while the curvature and torsion are kept constant. There is a large time/phase delay in the tracking of velocity with the feedforward control, which is expected due to the observed slow convergence speed of the open-loop system. Furthermore, there is significant tracking error with the PI control, which is in contrast to the case of tracking fixed references as shown in Figure 12.14. With the 2-DOF controller, the tracking performances are significantly improved in terms of the time/phase delay and the tracking error. It is also observed from Figure 12.17a and 12.17b that the variables fluctuate even though the reference is a constant. This indicates the strong coupling among control inputs on the spiral geometric features.

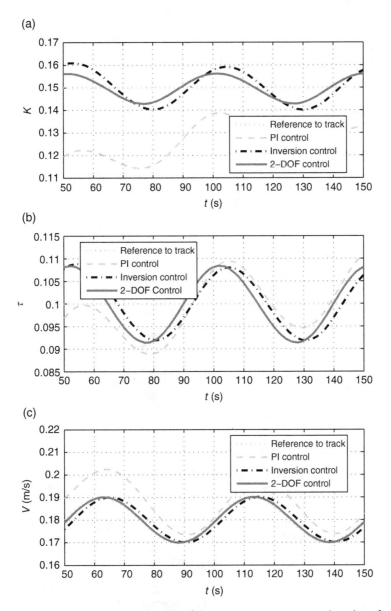

Figure 12.17 The simulation results of the geometric parameters when the reference velocity changes as a sinusoid function with respect to time while curvature and torsion are kept constant. (a) Curvature; (b) torsion; (c) velocity.

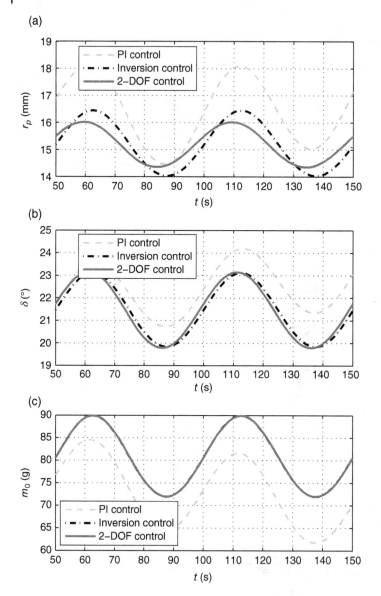

Figure 12.18 The simulation results of control inputs on when the reference velocity changes as a sinusoid function with respect to time while curvature and torsion are kept constant. (a) Displacement of movable mass; (b) rudder angle; (c) net buoyancy.

Exercises

1 Consider the reduced-order longitudinal model of the underwater glider that has an elevator control surface (12.20). Prove the exponential stability of the equilibrium point at the origin using the Lyapunov function given by (12.21).

2 Assume an underwater glider with an elevator control surface has the same parameters as in Section 12.2.3. Simulate the closed-loop system using the reduced-order dynamic model (12.20) with passivity-based controller (12.27) and (12.28). Compare the simulation results with that for the full-order closed-loop system (Section 12.2.3). Comment on the performance difference and discuss why the controller designed for the reduced-order system is able to stabilize the full-order system as well.

3 Consider the reduced-order dynamic model (12.20) for the gliding motion in the longitudinal plane. The system parameters are the same as in Section 12.2.3. The control objective is to regulate the gliding angle to any given set point. Design a sliding model controller and simulate the controlled gliding path. Study through simulation and comment on the influence of the controller parameters (e.g., k_1, k_2, and k_3 as in (12.55) on the convergence speed and steady-state error).

4 Consider an underwater glider that has a rudder control surface with the system parameters as follows (the same as in Exercise 2 in Chapter 11).

$$C_{D_0} = 0.5 \qquad C_D^\alpha = 15 \text{ rad}^{-2} \qquad C_{L_0} = 0.1$$

$$C_L^\alpha = 20 \text{ rad}^{-1} \qquad C_{M_0} = 0 \text{ m} \qquad C_{M_P}^\alpha = 1 \text{ m/rad}$$

$$S = 0.1 \text{ m}^2 \qquad m_w = 10 \text{ kg} \qquad r_w = 5 \text{ cm}$$

$$m_1 = 10 \text{ kg} \qquad m_3 = 15 \text{ kg} \qquad \bar{m} = 2 \text{ kg}$$

$$C_{F_S}^\beta = -2 \text{ rad}^{-1} \qquad C_{M_R}^\beta = -0.3 \text{ m/rad} \qquad C_{M_Y}^\beta = 0.5 \text{ m/rad}$$

$$K_{q1} = -0.1 \text{ m} \cdot \text{s/rad} \qquad K_{q2} = -0.5 \text{ m} \cdot \text{s/rad} \qquad K_{q3} = -0.1 \text{ m} \cdot \text{s/rad}$$

$$C_{F_S}^\delta = 1.5 \text{ rad}^{-1} \qquad C_{M_Y}^\delta = -0.2 \text{ m/rad} \qquad m_2 = 10 \text{ kg}$$

$$J_1 = 5 \text{ kg} \cdot \text{m}^2 \qquad J_2 = 0.5 \text{ kg} \cdot \text{m}^2 \qquad J_3 = 0.5 \text{ kg} \cdot \text{m}^2$$

Calculate the linearized model of the spiral dynamics (12.56) evaluated at control inputs $m_0 = 0.5$ kg, $r_p = 10$ cm, and $\delta = 30°$. Design an H_∞ controller to stabilize the steady spiral motion and simulate the closed-loop linearized dynamics.

References

Bhatta, P. and Leonard, N. E., "Nonlinear Gliding Stability and Control for Vehicles with Hydrodynamic Forcing," *Automatica*, Vol. **44**, No. 5 (2008), pp. 1240–1250.

Kelley, C. T., *Solving Nonlinear Equations with Newton's Method*, Society for Industrial Mathematics, 2003.

Khalil, H. K., *Nonlinear Systems*, Prentice Hall, Upper Saddle River, NJ, 2002.

Pressley, A. N., *Elementary Differential Geometry*, Springer, 2010.

Warner, F. W., *Foundations of Differentiable Manifolds and Lie Groups*, Vol. **94**, Springer, 1983.

Zhang, F. and Tan, X., "Three-Dimensional Spiral Tracking Control for Gliding Robotic Fish," *Proceedings of the 53rd IEEE Conference on Decision and Control* (Los Angeles, CA, 2014, accepted).

Zhang, F. and Tan, X., "Passivity-Based Stabilization of Underwater Gliders with a Control Surface," *Journal of Dynamic Systems, Measurement, and Control*, Vol. **137**, No. 6 (2015), 061006.

Zhang, F., Ennasr, O., and Tan, X., "Gliding Robotic Fish: An Underwater Sensing Platform and Its Spiral-Based Tracking in 3D Space," *Marine Technology Society Journal*, Vol. **51**, No. 5 (2017), pp. 71–78.

Zhang, S., Yu, J., Zhang, A., and Zhang, F., "Spiraling Motion of Underwater Gliders: Modeling, Analysis, and Experimental Results," *Ocean Engineering*, Vol. **60** (2013), pp. 1–13.

Appendix A

Demonstrations of Undergraduate Student Robotic Projects

A.1 Introduction

The appendix is devoted to describing two robotic projects conducted by students in the Department of Electrical and Computer Engineering at George Mason University. In addition to the descriptions, links to videos of the projects are also included.

A.2 Demonstration of the GEONAVOD Robot

To view the video on the GEONAVOD project go to the book webpage at www. wiley.com or http://www.video.youtube.com/watch?v=IENfst4Dcg8. In this video a small mobile robot can be seen carrying out a particular task. The following describes the robot and the scenario represented in this demonstration.

A fence-like structure was built to support three ultrasound transmitters as well as a radio transmitter approximately 2 m above the floor where the robot operated. These transmitters are coordinated and synchronized via signals from the radio transmitter. Initially, a radio signal is transmitted to synchronize the clock on the robot with the radio transmitter clock. Then in a predetermined sequence at predetermined time intervals, the three ultrasound transmitters transmit. The receiver on the robot determines the time of arrival of each transmission, and knowing the schedule for initiation of each transmission, it computes the time of travel from each transmitter to the robot. Knowing the speed of travel of the ultrasound signal, the distance of the robot from each transmitter is determined. Then using an iterative process, the robot coordinates are computed. The dimension of the problem is reduced to two since the z component is known, i.e., the robot is on the floor.

In fact the robot has two ultrasound receivers, one at each end, so this process is carried out for each of these. With the determination of position of each end of the robot, heading as well as location can be determined. The task of the robot

Mobile Robots: Navigation, Control and Sensing, Surface Robots and AUVs,
Second Edition. Gerald Cook and Feitian Zhang.
© 2020 by The Institute of Electrical and Electronics Engineers, Inc.
Published 2020 by John Wiley & Sons, Inc.

in the demonstration is to navigate along a path with three way-points. The robot periodically recalculates its position and heading as it traverses this path. In the demonstration, a team member intervenes from time to time to set the robot off course and test its ability to right itself and get headed back toward the next way-point.

As the robot travels it searches for signals of interest. In this case, it is looking for infrared sources. The search is carried out via a rotating turret with an IR detector mounted on top of the robot. Once an IR source is detected, the robot stops and the angle of the turret at detection is noted. The robot then turns the appropriate amount to face the source of the signal. A laser sensor is then used to measure the distance to this source and the coordinates of the source are computed. The robot then navigates its way to this object of interest.

The design team was comprised of Edward Smith, Charles Purvis, Johnny Garcia, and Ulan Bekishov. The design, construction, assembly, and testing were entirely accomplished by this team who at the time were all undergraduate students in the Department of Electrical and Computer Engineering of George Mason University.

A.3 Demonstration of the Automatic Balancing Robotic Bicycle (ABRB)

To view the video on the ABRB project go to the book webpage at www.wiley. com or http://www.youtube.com/watch?v=vczwea6iv_M&feature=related. In this video, the automatic balancing robotic bicycle can be seen demonstrating its capability. A potential application of this robot would be for the transfer of materials through spaces that are too tight for a four-wheeled vehicle. The following describes the robot.

The robotic bicycle balances itself through the use of a reaction wheel mounted orthogonal to the longitudinal axis and powered by an electric motor. Whenever the motor accelerates the reaction wheel, a torque is created within the motor and transferred to the bicycle frame on which it is mounted. Sensors measure the roll angle of the bicycle and compare this value to vertical. Whenever a deviation in the roll angle from vertical is detected, a controller directs the motor to accelerate the reaction wheel. The resulting torque acts to rotate the bicycle back toward vertical at which time the motor torque can be set to zero. The balancing principle is similar to that of controlling an inverted pendulum without moving the supporting base, and two members of the design team had previously successfully applied a reaction wheel to this problem.

Several variables are measured and fed into the controller as feedback signals. These include bicycle roll angle, roll angle rate, and reaction wheel rate. The first two signals are required to stabilize the bicycle in the vertical position while the last one is needed to bring the rotation velocity of the reaction wheel back to zero, keeping it from being driven into saturation. The equations of motion based on first principals were derived and the resulting model was considered with regard to controller design; however, the final values of the controller gains were determined empirically through experimentation with the actual bicycle.

The locomotion of the bicycle is provided by another small electric motor, connected to the rear wheel with a belt drive. This drive motor is controlled externally by a human via a wireless handheld remote controller. Sensors for collision avoidance are incorporated into the system and can override the locomotion commands if needed. Power for all the operations is provided by onboard batteries.

The design team was comprised of Aamer Almujahed, Jason Deweese, Linh Duong, and Joel Potter. The design, construction, assembly, and testing were accomplished entirely by this team who at the time were all undergraduate students in the Department of Electrical and Computer Engineering of George Mason University.

Index

Mobile Robots: Navigation, Control and Sensing, Surface Robots and AUVs,
Second Edition. Gerald Cook and Feitian Zhang.
© 2020 by The Institute of Electrical and Electronics Engineers, Inc.
Published 2020 by John Wiley & Sons, Inc.

CPSIA information can be obtained
at www.ICGtesting.com
Printed in the USA
LVHW080803140422
716182LV00004B/91